06576729

JUANITA HIGH SCHOOL
10601 N.E. 132nd St.
Kirkland, WA   98034

United States v. Virginia

Virginia Military Institute Accepts Women

Barbara Long

Landmark Supreme Court Cases

Enslow Publishers, Inc.

40 Industrial Road   PO Box 38
Box 398   Aldershot
Berkeley Heights, NJ 07922   Hants GU12 6BP
USA   UK

http://www.enslow.com

Library of Congress Cataloging-in-Publication Data

Long, Barbara, 1954–
    United States v. Virginia : Virginia Military Institute accepts women
/ Barbara Long.
        p. cm. — (Landmark Supreme Court cases)
Includes bibliographical references and index.
Summary: Presents information about the Supreme Court case that questioned the Virginia
Military Institute's male-only policy and also refueled the debate regarding
private, single-gender schools.
    ISBN 0-7660-1342-1
    1. Virginia Military Institute—Trials, litigation, etc.—Juvenile
literature. 2. Sex discrimination in education—Law and legislation
United States—Juvenile literature. 3. Military education—Law and
legislation—United States—Juvenile literature. 4. Women—
Education—Law and legislation—United States. [1. Virginia
Military Institute—Trials, litigation, etc. 2. Sex discrimination in
education—Law and legislation. 3. Military education—Law and
legislation. 4. Women—Education—Law and legislation.] I. Title:
United States versus Virginia. II. Title. III. Series.
    KF228.V57 L66   2000
    344.73'0798—dc21
                                                                    99-050671

Printed in the United States of America

10 9 8 7 6 5 4 3 2 1

To Our Readers:
All Internet addresses in this book were active and appropriate when we went to press. Any
comments or suggestions can be sent by e-mail to Comments@enslow.com or to the address
on the back cover.

Photo Credits: Courtesy of the Office of Public Relations, Virginia Military
Institute, Lexington, VA, pp. 7, 11, 14, 66, 105; Courtesy of the Photo
Department, The Citadel, Charleston, SC, pp. 36, 47, 53, 91, 93, 96, 98; The
Library of Virginia, Richmond, VA, p. 32; Supreme Court Historical Society,
United States Supreme Court, Washington, D.C., pp. 26, 40, 58, 75, 77.

Cover Photo: Cadets at VMI in formation, Courtesy of the Office of Public
Relations, Virginia Military Institute, Lexington, VA

Contents

Acknowledgments

The author acknowledges and sincerely thanks Michael M. Strickler, public relations director at VMI, and Russell K. Pace at The Citadel photo department for their cooperation in acquiring material and photographs for this book.

Dedication

To my parents, William H. and Jean J. Long, who have always encouraged and supported their daughters, as General Thomas Jackson did his cadets, to be "whatever you resolve to be."

1

Welcome to the "Rat" World

Entering the Virginia Military Institute (VMI), located in Lexington, Virginia, is not a typical "welcome to college" experience. Rather, it is "welcome to the 'rat' world." First-year students not only say good-bye to their families, but also good-bye to their civilian, or street, clothes, and hello to military uniforms. They also say good-bye to their individual hairstyles and hello to shaved heads. First-year students are referred to as "rats," and during their first year at school they must live under the "rat system." The rat system applies to every new cadet, regardless of background or earlier education. The six-month initiation, which is conducted by the upper-class cadets, is very much like that of a United States military boot camp.

Rats must always walk at attention and keep their chins tucked into their necks, a position known as bracing. They must run double-time on all stairs. First-year cadets may speak to upper-class cadets only when those students speak to them. The new recruits may not leave the campus at night, and they cannot have an entire weekend off until their second term of school. Upper-class cadets often hold "sweat parties," making rats exercise until they are exhausted. The older cadets may also conduct detailed inspections of the new cadets' rooms and uniforms. Rats must listen to frequent and loud yelling in their ears.[1]

A small break in rules can result in a rat having to do pushups, being confined, or marching on a "tour" around campus. More serious offenses, such as violating the cadet honor code, can result in a cadet's being expelled from the school. The honor code states, "A cadet will not lie, cheat or steal, or tolerate those who do." The code is contained in the sixty-four-page booklet that serves as the VMI bible, which new students are expected to memorize.[2] Every minute of every day—from reveille, or morning wake-up call, at 6:55 A.M. until lights out at midnight—the cadets live and study within the rat system. There is little escape from the mental and physical stress, and there is hardly any privacy.

Cadet Life

All VMI cadets live in the Barracks, as those before them have done since the 1850s. This stark four-story building has no telephones and no televisions. Doors to the cadets' rooms have no locks, and the windows simply have shades that may be pulled down only when a cadet is dressing or undressing. Given these conditions, it is not surprising that nearly 10 percent of

The stark four-story building, with no telephones and no televisions, known as the Barracks is shown here. All VMI cadets live in the Barracks.

the first-year cadets drop out of VMI in the first six weeks.[3] After one year, an average of 22 percent of the freshman class decides to leave VMI.[4] It is not the typical life most students imagine when they think of going to college.

The VMI experience is a life-changing one. A similar experience occurs at only one other college in the United States. That institution is The Citadel, founded in 1842, in Charleston, South Carolina. This military college also follows the tough educational philosophy of VMI. At The Citadel, first-year students are called "knobs," but their daily routine is very much like that of the VMI rats.

Why would anyone want to enter such an intensely stressful program? And how have these seemingly old-fashioned institutions survived into the twenty-first century?

One answer to these questions is that both these schools have a long and successful tradition of graduating outstanding leaders in many areas of life. About 18 percent of VMI graduates make the military their lifelong career.[5] Many of the rest of the school's graduates become successful business, professional, and government leaders. Such a distinguished tradition began long ago in the nineteenth century.

The History of VMI

In 1839 VMI was founded as an all-male educational institution. Today it is the oldest state-supported military college in the country. VMI's mission was, and is, to provide a thorough education within a framework of military discipline. In other words, the school's goal is to graduate citizen-soldiers. And the school has delivered on its educational promise for more than one hundred sixty years.

VMI graduates have fought in every American conflict since the Mexican War (1846–1848). During the Civil War—on May 15, 1864—247 VMI cadets joined Confederate forces led by General John C. Breckinridge. The cadets rallied to fight the federal troops heading into Virginia's Shenandoah Valley. On that bloody day, known as the Battle of New Market, ten VMI students were killed and many more were wounded. Still, the young cadets were instrumental in helping to defeat the seasoned Union soldiers. This conflict marked the only time in America's history that an entire student body has fought as a unit in battle. Federal troops burned VMI that year, but the institution was back teaching students in 1865, one year later.

Graduates of VMI include Thomas "Stonewall" Jackson, the well-known Confederate general who fought for the South so valiantly during the Civil War, and later became a member of the VMI faculty. Another alumnus was George C. Marshall, who graduated in 1901. Marshall, a five-star general, served as Army Chief of Staff during World War II. After the war, he created the European Recovery Program known as the Marshall Plan. This program offered economic and technical assistance from the United States to sixteen European countries after World War II. The program's goal was to help restore the economy of Western Europe and encourage economic growth and trade among the major countries that were not communist. For this plan, Marshall won the Nobel Peace Prize in 1953—the only soldier to do so.

No matter how VMI's educational approach is graded, it receives exceptionally high marks. The college truly inspires students to reach their fullest potential. In general, many cadets who apply to and are accepted at VMI are quite average. They have average abilities, average backgrounds, and average economic means.[6] Yet according to a 1996 survey, VMI ranked thirtieth among the nation's colleges and universities in the number of graduates occupying leadership positions in American business.[7]

Based on VMI's record, serious students looking for a quality education at an affordable price would do well to consider this military college. However, students

This statue shows George C. Marshall, a 1901 VMI graduate who went on to become a five-star general and Army Chief of Staff during World War II. The Marshall Plan, created by and named for George Marshall, offered economic help from the United States to sixteen European countries after World War II. Marshall won the Nobel Peace Prize in 1953 for his work on the Marshall Plan.

must meet the challenges of the rat system. The system clearly is not for everyone. Some are not up to the physical and mental challenges.

The VMI approach to educating young people is known as the adversative approach. It is based on the old English public-school concept. For years, English public schools employed the philosophy that mental and physical challenges help build character and increase achievement. The purpose of VMI's grueling system is to take away each cadet's sense of individuality in order to get students to bond through their shared hardships. The program builds unity as well as responsibility. The friendships formed in this demanding environment often last a lifetime. The adversative system also tests a person's ability to deal with continual stress. After completing their first year, cadets have an unshakable sense of self-confidence.

"It gives you the feeling that you will confront nothing in life that you can't handle," said Joe Bowers, who graduated from The Citadel in 1963. "You will not freeze, you will not choke up, you will do what has to be done."[8]

For more than one hundred fifty years, VMI's admissions policy of accepting only males went unchallenged. However, it was only a matter of time before some women would want to be a part of VMI's rich tradition. This was especially true in light of the civil rights movement of the 1960s and the women's movement of the 1970s. These movements, as you will learn in the next chapter, involved social and political changes in American society and brought about more equal treatment for African Americans and women. African Americans began to be admitted into VMI in the mid-1960s and now make up 7 to 8 percent of the student body.[9] Women waited much longer for admission to VMI. By the 1990s some women felt as prepared to endure the "no pain, no gain" system as their male peers did. However, women's applications were routinely rejected—solely on the basis of gender— at both VMI and The Citadel. On VMI's campus, Jackson Arch, an entryway to the school, is etched with the famous general's belief: "You may be whatever you resolve to be." Does this philosophy apply to men only?

Legal Challenges

In 1990 the United States Department of Justice filed a lawsuit against VMI on behalf of a Virginia woman who wanted to enter the school. The Department of Justice claimed that VMI's male-only admissions policy violated the Civil Rights Act of 1964 and the Equal Protection Clause of the Fourteenth Amendment of the United States Constitution.

Jackson Arch, named for Thomas "Stonewall" Jackson, is an entryway to the campus of VMI. Etched on the arch is the following belief of General Jackson: "You may be whatever you resolve to be."

In 1993, The Citadel received and accepted an application from a nineteen-year-old high school senior named Shannon Faulkner. However, as soon as school officials realized that Shannon was female, her application was rejected. She filed a lawsuit against the school that would take over two years to be decided.

Should women be allowed to attend these traditionally all-male public schools? Would women destroy the educational systems of the colleges? In the mid-1990s the United States Supreme Court had to answer these questions. The case was sent to the nation's highest court because these all-male schools refused to admit female cadets, even when the young women met the school's admissions requirements. The Supreme Court would have to decide if the schools' admissions policies violated the Equal Protection Clause of the Fourteenth Amendment to the United States Constitution.

2

The Start of a Long March

The United States was formed because most people living in the thirteen British colonies believed that certain freedoms were a right of every human being. Most people in Colonial America believed that they should be able to govern themselves rather than be ruled by a king or queen. With the Declaration of Independence, written by representatives of the people and adopted on July 4, 1776, the colonists cut ties with Great Britain and started the American Revolution, also known as the Revolutionary War. After the Revolutionary War, leaders of the independent United States of America met to plan how to govern the new nation. They believed that citizens of the United States should have certain basic rights, such as freedom of speech, freedom to bear arms, and freedom to worship as they pleased. With these freedoms in mind, the Constitution of the United States was written.

In addition to outlining citizens' rights, the United States Constitution serves as the country's primary law. It acts as a guide for the government as well as for the principles that the government uses to rule. The words of Supreme Court Chief Justice John Marshall in the early nineteenth century are an excellent summary of the purpose of this legal document. The Constitution was "intended to endure for ages to come, and, consequently, to be adapted to the various crises of human affairs."[1]

The first ten amendments to the United States Constitution are called the Bill of Rights. These amendments provide basic legal protection of individual rights. Over the years, other amendments have been added to the Constitution. The Fourteenth Amendment deals with "equal protection under the law" for all people. This amendment, which was adopted in 1868, has often been used to ban discrimination and separate policies, called segregation, based on race and gender.

The Civil Rights Movement

One of the most well-known United States Supreme Court decisions dealing with the Fourteenth Amendment is the 1954 case *Brown* v. *Board of Education of Topeka, Kansas.* Before *Brown* v. *Board of Education,* schools were segregated by race. In 1896 the Supreme Court had decided, in the case of *Plessy* v. *Ferguson,* that racial segregation was legal as long as equal facilities were available for both whites and blacks. This thinking became known as the "separate but equal" doctrine. *Plessy* v. *Ferguson* involved passenger accommodations on railroad lines. However, in the following years, the Court's "separate but equal" doctrine was applied to many areas of public life in America. The Supreme Court did not start to demand equal treatment of all races until the 1940s. In the meantime, many white and black students attended separate schools, especially in the South.

When Linda Brown was refused admission to an elementary school in Topeka, Kansas, simply because she was African American, her parents filed a lawsuit. Similar suits from South Carolina, Virginia, and Delaware were included in the *Brown* case. Attorneys for the African-American students argued that the segregated schools were not equal and could not be made equal. The individual lawsuits boiled down to the following question: Does the Equal Protection Clause of the Fourteenth Amendment prohibit racial segregation in public schools?

On May 17, 1954, all nine Supreme Court Justices

agreed that the "separate but equal" doctrine as applied to segregated schools violated the Equal Protection Clause of the Fourteenth Amendment. Chief Justice Earl Warren wrote the opinion. In it he stated that "separate education facilities are inherently [by nature] unequal."[2] He stated that separating children in the schools for reasons based only on race makes some children feel less equal than others. Warren believed that this kind of unfair treatment could affect the children throughout their lives.

However, the Court's landmark decision did not overturn all segregation in the nation's schools. On May 31, 1955, the Supreme Court directed local courts to require "a prompt and reasonable start toward full compliance" with the decision to ban racial segregation in American schools.[3] Many schools began to integrate white and black students in the same buildings. This was often accomplished by students being bussed from one neighborhood school to another. Although busing helped integrate many schools, the policy was controversial, constantly under challenge. Busing to achieve integration has decreased dramatically in recent years.

The Court's decision was the basis for integrating many public schools and banning public policies that permitted segregation. The *Brown* decision also had a great effect on the civil rights movement in the 1960s.

Indeed, the *Brown* decision greatly impacted American society, serving as a powerful weapon in the battle against race discrimination.

The Women's Rights Movement

The demand for equal rights for American women began about the same time as did the demand to end slavery within the United States. In fact, many women became active in the abolition movement—the movement to end slavery. In February 1838, Angelina Grimké spoke before state lawmakers in Massachusetts, presenting an antislavery petition signed by twenty thousand women.[4] She was the first woman ever to speak before lawmakers in the United States, and her action marked the end of women's public silence.[5]

The debate about the rights of African Americans also raised questions about the rights of women. At that time in the United States, there were no laws protecting women's rights. Concerned about women's status, several women held the Women's Rights Convention on July 19 and 20, 1848. Elizabeth Cady Stanton, one of the group's leaders, suggested creating a list of demands for women, based on the Declaration of Independence.[6] During the two-day meeting, the Declaration of Sentiments was written and twelve resolutions were added to it. The resolutions included demands that women have the right to own property, exercise free speech, obtain divorce, achieve equal educational opportunities, and—most important—vote. This convention, held in Seneca Falls, New York, moved women's rights from an idea to a goal.[7] Yet the goal was not easily achieved. On one hand, many lawmakers who considered the women's suffrage movement a joke refused to pass laws to protect women's rights. On the other hand, many people, both men and women, felt threatened by feminism (the belief in equal treatment of men and women).

After the Civil War, the Fourteenth Amendment was passed, giving equal protection under the law to all citizens, including African Americans. However, when it came to the issue of representation in voting, the word *male* was inserted into the law, specifically indicating "male citizens." For the first time, the United States Constitution contained the word *male* rather than *people* or *citizens*.[8] As noted earlier, the Fourteenth Amendment was passed in 1868. Yet the Nineteenth Amendment, which granted women the right to vote, was not adopted until 1920. American women had to struggle and protest for just over a half century more before they could exercise their right to vote.

In the 1970s, feminists once again became vocal, protesting discrimination based on gender. Feminists demanded equal pay for equal work, and they demanded that the educational and career opportunities available to men also be available to women. The Equal Rights Amendment (ERA), intended to ban sexism, was proposed as the Twenty-Seventh Amendment to the United States Constitution. This proposal was originally drafted by Alice Paul of the National Woman's Party and introduced in Congress in 1923 during the suffrage movement. The government did not act on Paul's proposal until the 1970s, when a wider public—largely fueled by the National Organization of Women (NOW)—began to support women's equal rights and reintroduced the ERA in Congress. The main purpose of the modern ERA was to give women constitutional protection not ensured by the Equal Protection Clause of the Fourteenth Amendment. Feminists wanted to make gender discrimination unconstitutional, knowing that violation of constitutional rights is considered a serious offense in the legal system of the United States.

Congress approved the ERA in 1971, and the Senate followed suit one year later. On May 22, 1972, the amendment was submitted to the states for ratification. The deadline for ratification was first set for March 1979, but in 1978 the deadline was extended to 1982. On June 30, 1982, ratification of the ERA was just three states short of the thirty-eight states needed for the amendment to become part of the Constitution. On July 14, 1982, the ERA was again reintroduced to Congress, but was defeated in the House on November 15, 1983. Despite a great deal of effort, this amendment did not become law. However, women's voices have been heard, and many changes have been made to their status in society. Women have become more visible in a variety of careers and positions, serving as inspirational role models.

Equal Opportunity and Equal Responsibility

In their march for equal opportunity and equal responsibility, women began demanding the right to serve in the country's armed forces. Women also began to push for admission to the four United States military academies. These institutions are the United States Military Academy at West Point, New York; the United States Naval Academy at Annapolis, Maryland; the Air Force Academy near Colorado Springs, Colorado; and the Coast Guard Academy in New London, Connecticut. Women first entered classes at these academies in 1976, graduating in 1980. By 1999, women made up 15 percent of the graduating class at the United States Naval Academy. The top graduate that year was a woman—Mary Godfrey of Doresman,

Wisconsin. Class standing is determined not only by grades but by military ability and conduct as well.[9]

Prior to these milestones, in 1973, North Georgia College—a public school located in Dahlonega, Georgia—began enrolling women as cadets in its military program. Many women said that they joined the corps to prepare for military careers and that they valued the discipline the college offers.[10] Women were, and are, also attracted to North Georgia because of the scholarships that the state offers students who are enrolled in military programs.[11] In the decades that followed their admission, women have risen to top leadership positions in the cadet corps and joined previously all-male athletic teams. By 1990, male and female cadets were living in the same dormitory, and no one spent much time thinking about the gender integration of the corps.

In 1982 the United States Supreme Court decided the case of *Mississippi University for Women* v. *Hogan*.[12] This case involved "reverse discrimination," or bias against a male or a white person rather than a female or a minority person. A young man named Joe Hogan wanted to enter the school's nursing program so he would not have to drive seventy-five miles to a coed nursing program. The university was a public, not a private school. However, he was refused admission because the program was for women only. Justice Sandra Day O'Connor's opinion for the Court held that the Mississippi University for Women could not exclude men from its nursing school because the college did not have valid reasons for its discriminatory female-only admissions policy.

Thus, the university's admissions policy was in violation of the Fourteenth Amendment's Equal Protection Clause. Justice O'Connor added that the admissions policy tended to preserve and promote the stereotype of nursing as strictly women's work. The Court's ruling set constitutional standards for reviewing the admissions policies of single-gender public colleges and universities. Do all-male or all-female schools have a valid reason for existing? According to the Supreme Court, any distinction of individuals on the basis of gender must have valid reasons and must serve important government goals. The Court's decision did leave open the possibility that states could offer single-gender schools in cases where such programs would actually promote equality.[13]

The Struggle for Equality

The struggle for equality was an important issue during most of the twentieth century. And it was a big reason for the fact that in the early 1990s, women began to

Justice Sandra Day O'Connor wrote the Supreme Court's 1982 *Mississippi University for Women* v. *Hogan* decision. This case involved "reverse discrimination," or bias against a male or a white person rather than a female or a minority person.

question the male-only admissions policies of the nation's two state-supported military academies—the Virginia Military Institute in Lexington, Virginia, and The Citadel in Charleston, South Carolina.

Challenges to the two schools' admissions policies were natural, given the civil rights and the women's rights movements of the recent past as well as the social climate of the country near the end of the twentieth century. In a time when men and women competed equally in many areas of society, why did VMI and The Citadel continue to refuse to admit women to their schools?

Some people believed that VMI and The Citadel were discriminating on the basis of gender. They felt that because both schools were public educational institutions funded by state dollars, both should be in compliance with the United States Constitution's Fourteenth Amendment.

Other people believed that the same rights and opportunities should not be available to both men and women. They thought that there were good reasons for having separate policies and schools based on gender. They believed that women had no place at VMI and The Citadel because young women could get training at other colleges. These people were concerned that women would destroy both schools' adversative approach to education, which had proved so successful over the years.

The long march for freedom and equality, started in the late 1700s, was not yet over. Cases of unfair treatment and discrimination still occurred in the 1990s, and they were being challenged. Could VMI and The Citadel continue to refuse to admit female applicants simply on the basis of gender? Could these all-male military colleges do so under the United States Constitution?

3

The Legal Battle

At the end of the 1980s, the Virginia Military Institute and The Citadel were still refusing to admit female applicants to their schools. Then, in 1989, an unidentified young woman from Virginia wrote a letter to the United States Department of Justice. In her letter, the woman complained about not being able to enter VMI because of its all-male tradition. In January 1990, Richard Thornburgh—United States attorney general under President George Bush—sent letters to Virginia's newly elected governor, Douglas Wilder. Thornburgh addressed his letters to Wilder because VMI is a state school, and as governor, Douglas Wilder represents the interests of the state. In fact, at that time, VMI was receiving $9 million each year from the state. This large amount of money was nearly a third of the school's budget.[1] Furthermore, the governor of Virginia had, and still has, the responsibility of appointing members to the school's board. These facts make the relationship between VMI and the state very clear. VMI, indeed, is a public school funded by the state of Virginia. In his letters, Attorney General Thornburgh ordered VMI to admit women or be sued by the Department of Justice for violating the Fourteenth Amendment of the Constitution.

First Shots Fired

Governor Wilder took the gender discrimination issue to VMI's Board of Visitors, which helps govern the college. He wanted to let the school's leaders decide whether to accept or fight the order from the United States government. The VMI Board of Visitors voted to fight the order to admit women to the military school. Virginia's female attorney general, Mary Sue Terry, agreed that VMI's all-male tradition was worthy of protecting and decided to defend the school's decision.

Terry, the VMI Board of Visitors, and the VMI Foundation (an organization that raises funds for the college) filed two lawsuits of their own. In their lawsuits, filed on February 5, 1990, the three parties argued that VMI was a necessary and distinctive part of the state's diverse system of higher education. They also asserted that "women were not denied equal opportunity by VMI's admissions policy" because of the coed military corps of cadets located at nearby Virginia Polytechnic Institute.[2] The parties also argued that "the forced admission of women [would] eliminate the option of a unique undergraduate experience for men, while providing no increased educational opportunities for women."[3]

In March 1990, the United States government sued Virginia and VMI in the United States District Court for the Western District of Virginia on behalf of the unidentified Virginia woman who had been refused admission to the military college. The lawsuit charged that the VMI male-only admissions policy violated the Civil Rights Act of 1964 because the school received state funds and openly discriminated against women. The United States government also pointed out that VMI's all-male tradition violated the Equal Protection Clause of the Fourteenth Amendment. The case received a great deal of attention in Virginia for two reasons. First, VMI graduates form an important network throughout the state. Second, the lawsuit put Governor Wilder and Attorney General Terry of Virginia in the delicate position of having to decide which group of voters to support.[4]

VMI Wins First Round

In 1991 the United States Court of Appeals for the Fourth Circuit upheld the District Court's decision and ruled, two to one, in favor of VMI's admissions policy. District Court Judge Jackson L. Kiser said that VMI's program and policy "promotes diversity" within Virginia's higher-education system.[5] He believed that

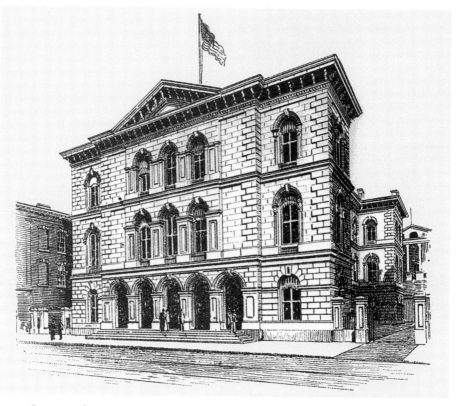

In 1991 the United States Court of Appeals for the Fourth Circuit ruled in favor of VMI's all-male admissions policy.

VMI offered a unique educational opportunity not found at other Virginia colleges. He also stated that admitting women to VMI would harm the school by forcing it to change its basic philosophy, and therefore, most likely ruin the quality of education provided by the college. Kiser himself had received his law degree from an all-male law school near the VMI campus. He had attended Washington and Lee University thirty-three years before it became coed.[6]

To reach his decision, Judge Kiser cited the Supreme Court's 1982 decision in *Mississippi University for Women* v. *Hogan*, mentioned earlier. In that case, the Supreme Court had ruled that any distinction of individuals on the basis of gender had to serve important government goals.[7] Judge Kiser wrote in his decision that VMI met this condition. He said that VMI was right in trying to preserve the "adversative model" of education that is based on "physical rigor, mental stress, absolute equality of treatment, absence of privacy, minute regulations of behavior, and indoctrination."[8]

Judge Kiser also cited the studies of three well-known educators who researched the effects of single-gender education. He believed there was strong evidence to support VMI's claim that some students, both male and female, benefit from attending a single-gender college.[9] However, one of these experts—Alexander W. Astin, director of the Higher Education Research Institute at the University of California at Los Angeles—claimed that the data Kiser used to make his decision were out of date. Astin said that his book *Four Critical Years*, which Kiser referred to in his decision, was based on research that he conducted in the late 1960s and 1970s.[10]

The two other education researchers—Richard C. Richardson, Jr., a professor of educational leadership and policy at Arizona State University, and David Riesman, a professor of sociology at Harvard University and an author of many books about higher education—were paid to serve as expert witnesses for VMI. Richardson said that many people were surprised by his support of VMI because he often challenged various schools to enroll more minority students. He believed that his support was not different from his usual views. "I've always said that one does not sacrifice quality to achieve diversity."[11] Riesman said he was happy to be involved in the case because he had long believed in single-gender education.[12]

After Judge Kiser's ruling, Robert H. Patterson, Jr., the chief lawyer for VMI, declared the decision "a victory for single-sex education, educational diversity, and common sense."[13] He then urged the United States Department of Justice to accept this decision and not appeal it. The government had two months to appeal the court's decision. Lieutenant General Claudius E. Watts, III, president of The Citadel at that time, echoed Patterson's pleasure about the court's ruling. Watts said, "A very dark cloud has been lifted from VMI and The Citadel."[14]

Those on the government's side heartily disagreed. Marcia D. Greenberger, president of the National Women's Law Center, claimed that the ruling "missed the point altogether" since Judge Kiser "did not understand that gender discrimination hurts both men and women, in that it pigeonholes them in artificial ways that don't reflect individual abilities."[15] A senior counselor for the National Women's Law Center, Ellen Vargyas, made another point. "We're not talking about whether there is a role for single-sex education. The real question is, can the brother rats have male bonding with tax money from the state of Virginia?"[16] Kiser's ruling was made in 1991 during the time of Operation Desert Storm (a United States military attack against Iraq after Iraq invaded Kuwait), in which both female and male soldiers served together. The judge's thinking seemed to go against the current trend of integrating women into the military services.[17]

Lieutenant General Claudius E. Watts, III, president of The Citadel at the time of the appeals court ruling in 1991, applauded Judge Kiser's decision.

United States Government Wins
Second Round

In 1992, the United States Court of Appeals for the Fourth Circuit heard the government's appeal of Judge Kiser's ruling. The court agreed with VMI that opening the school to women would change the mission and philosophy of the military college. However, it did not agree that the all-male environment promoted diversity. The appeals court panel ruled that Virginia could not fund a male-only public institution without creating some type of alternative program for women. The panel agreed that VMI's male-only admissions policy would not be a legal problem if the school were private rather than public. The general counsel of the American Council of Education, Sheldon Steinbach, said the court's ruling should help officials of private single-gender schools understand that they were not in danger of having to integrate their facilities.[18] The court of appeals then gave VMI three choices: (1) admit women, (2) become a private institution, or (3) offer a similar program for women.

After this ruling, VMI quickly asked the Supreme Court to hear an appeal. Lawyers for the military college argued that the appeals court decision opened every all-female college to possible discrimination lawsuits. They pushed this idea further by saying that the ruling could lead to challenges to state support for shelters for battered women and to single-sex prisons.[19]

At the time, the country had four public colleges that stressed single-gender education. The two all-male public schools were VMI and The Citadel, and the mainly all-female public schools were the Mississippi University for Women and Texas Woman's University. The two women's colleges admitted men to some of their programs, but they generally focused on educating females. VMI was the school under the most pressure from the federal government to integrate men and women, although the Department of Justice was also investigating The Citadel's admissions practices.

The government lawsuit against VMI was still a political problem for many state leaders. Governor Wilder publicly disapproved of excluding women from the state military college.[20] Because of the governor's position, Attorney General Terry of Virginia removed herself from the case.[21] In addition, one former member of the Virginia state senate, Emilie F. Miller, believed that she lost her elected position because of her efforts to force VMI to admit women.[22]

In the early months of 1993, seven women's colleges filed two briefs that urged the Supreme Court to hear an appeal of the lower court's decision declaring VMI's admissions policy illegal. Saint Mary's College of North Carolina, Southern Virginia College for Women, Randolph-Macon Woman's College, as well as Mary Baldwin, Hollins, Sweet Briar, and Wells Colleges believed that the VMI case threatened their legal existence as well as that of the public military colleges. However, most people felt that these private women's colleges would not be affected by the court's ruling against VMI. They pointed out that the Department of Justice's case was based on the Equal Protection Clause of the Constitution, which applies only to government agencies. Congress had given specific exemptions to private undergraduate single-gender programs when it enacted civil rights laws.[23]

In May 1993, the United States Supreme Court refused to examine the appeals court's ruling that found VMI's admissions policy in violation of women's rights. However, Associate Justice Antonin Scalia offered an opinion that disagreed with the Court's refusal to hear an appeal of the case. He stated that the Court should consider the school's arguments before a national institution as time-honored as VMI is forced to change.[24] Associate Justice Clarence Thomas did not get involved in the discussion of the appeals since his son, Jamal, was a student there at that time.

Associate Justice Clarence Thomas did not participate in the Supreme Court's discussion of the appeal since his son, Jamal, was a student at VMI at that time.

VMI Counters

In 1992, the United States Court of Appeals for the Fourth Circuit in Richmond, Virginia, had asked VMI to come up with a solution to its discriminatory all-male admissions policy. Instead, VMI had appealed the ruling, hoping the Supreme Court would hear its case and allow the college to keep its all-male admissions policy. In 1993, when the Court decided not to hear the case, VMI was forced to find an acceptable solution to its discriminatory admissions policy. The federal district court was responsible for reviewing whatever option VMI chose.

As noted earlier, VMI's had three options. It could (1) admit women, (2) become a private institution, or (3) offer a similar program for women. Most school officials and cadets felt strongly about keeping women out of VMI. The second choice, making VMI a private school, was very expensive. In 1993, the state of Virginia provided $8.4 million to VMI, and South Carolina gave The Citadel $12.8 million.[25] Mary Sue Terry, the former attorney general of Virginia who was at that time the Democratic nominee for governor, issued a statement offering her opinion. She said:

> Given the current fiscal [economic] reality and the projected increase of 65,000 new Virginia students who will seek higher education in the next eight years,

41

building a new VMI or letting it go private do not appear to be viable [workable] options. It is time to look beyond additional costly litigation [lawsuits] and proceed in the fairest and most effective way—that is, to admit women.[26]

In the end, VMI decided to act on the third choice, to offer a similar institution or program for women.[27] In September the military college proposed that it be allowed to continue to admit only men and that it would help develop a separate leadership program for women. This program was called the Virginia Women's Institute for Leadership (VWIL) and would be held at Mary Baldwin College. The proposal set up a legal test of whether the "separate but equal" doctrine would hold up for VMI.

Virginia Women's Institute for Leadership

Mary Baldwin College is a private institution, located about thirty-five miles northeast of VMI. Cynthia H. Tyson, the president of the women's college at the time, said that the school had been trying to promote leadership education before the VMI case was filed.[28] Although students at the college had mixed views about VMI's plan, the faculty voted 52 to 8 to agree with the proposal.[29] President Tyson made it clear that Mary Baldwin was not endorsing VMI's approach to education or its legal case. "I do not have any influence on legal issues at all," she said. "The only influence the trustees have is on our own business, and we are here to work on behalf of advancing women."[30]

According to VMI's plan, women who enrolled in the VWIL program would live in separate dormitories, participate in leadership programs, and join the Reserve Officer Training Corps (ROTC). These cadets would not receive the adversative education that the VMI cadets received, since this experience is "conducive [favorable] to the development of confidence and self-esteem" for men but not for women.[31] To fund the leadership program, VMI offered $6.9 million, which provided about $10,000 for each student.[32]

In addition, the plan allowed administrators at Mary Baldwin College to control the admission of students to the VWIL program. The state of Virginia would give the college fifty-nine hundred dollars for each Virginia student admitted. This money was equal to the amount that the state provided for each VMI student. VMI believed its plan was a good solution to the problem of women who wanted to join the military college's all-male program.

Those in favor of forcing VMI to admit women, however, did not think that the VWIL program would work. These people claimed it would amount to unequal treatment of female cadets.[33] Ellen Vargyas, senior counselor for the National Women's Law Center, said she did not think it would be possible for Virginia to create an institution truly equal to VMI. She asked, "How in the world do you replicate [copy] an institution where a great deal of what it is, is several hundred years of history?"[34] She continued, "This would be a school without alumni, a school that is going to focus on mentoring [guiding] but doesn't have any mentors [guides]."[35] She further noted:

> . . . public institutions in Virginia and South Carolina had admitted black students only under federal pressure and with predictions from white political leaders that the quality of the colleges would be destroyed. The University of Virginia is doing just fine now. VMI will survive this.[36]

In April 1994, the district court approved VMI's proposal to create a women's leadership program, believing that VWIL met the requirements of the Equal Protection Clause of the Fourteenth Amendment. However, the United States Department of Justice appealed this ruling, claiming that the two programs— VMI and VWIL—were not and could not be equal.

In 1995 a panel of three judges of the United States Court of Appeals for the Fourth Circuit voted two to one to accept VMI's proposal, allowing that there were legitimate reasons for a state to operate single-gender colleges.[37] In the court's opinion, Judge Paul V. Niemeyer declared that teenagers are less distracted in an atmosphere in which there are no members of the opposite sex, especially "in an educational setting where the focus of learning is on matters other than relationships between the sexes."[38] The appeals court noted that it was not necessary for the two military programs to be exactly the same. It viewed both programs as providing the same opportunity for students to excel and become leaders.[39] However, the court wanted more information about the program. In particular, the panel wanted more information to prove that Virginia would recruit qualified managers for the women's program and that it would provide enough money to the program.

The dissenting judge on the panel, J. Dickson Phillips, Jr., disagreed with the majority of the appeals court. He believed that VMI was more concerned about protecting its all-male status than it was with providing women with a truly equal opportunity for education. He said that the state would satisfy the Constitution's equal protection requirement if it

> simultaneously opened single-gender undergraduate institutions having substantially comparative curricular and extra-curricular programs, funding, physical plant, administration and support services, and faculty and library resources.[40]

Phillips thought it evident that the proposed VWIL program, compared with VMI, fell "far short . . . from providing substantially equal tangible and intangible educational benefits to men and women."[41]

On the other hand, VMI's superintendent, Major General John W. Knapp, praised the court's decision. He said this ruling would allow VMI and Mary Baldwin to "put this litigious period [time of lawsuits] behind them and move forward with our fundamental goal of education."[42] However, the legal battle was not yet over. Marcia D. Greenberger, president of the National Women's Law Center, believed that the appeals court's decision turned the whole notion of equality on its head and that it relied on stereotyped notions of women.[43] As it did once before, the United States government filed an appeal.

The Citadel Fights the Same Legal Battle

During VMI's legal battle, women were also trying to gain admission to The Citadel. In 1993, Shannon Faulkner—a nineteen-year-old high school senior from Powdersville, South Carolina—applied to The Citadel. She was accepted until the school became aware that she was female. Then her application, which had omitted any references to her gender, was quickly

In 1993, Shannon Faulkner—a nineteen-year-old high school senior from South Carolina—applied to The Citadel. When the school became aware that she was female, her application (which had omitted any reference to her gender) was quickly rejected.

rejected. On March 2, 1993, Faulkner sued The Citadel, charging that the school's all-male admissions policy was unconstitutional.

On August 12, 1993, Judge C. Weston Houck of the United States District Court ruled that Faulkner could attend day classes until her lawsuit was settled. She was enrolled in nonmilitary programs only, and she could not live on the school campus or wear a cadet uniform. Faulkner's lawyers filed an appeal. On August 24—two days before Faulkner was to register for classes—the Court of Appeals for the Fourth Circuit stopped Judge Houck's order. Instead of being able to attend The Citadel, Faulkner spent the fall semester at the University of South Carolina in Spartanburg. Then on January 18, 1994, Shannon Faulkner became the first woman to attend day classes at The Citadel. However, this historic honor had its downsides. Many cadets hissed whenever Faulkner spoke in class. She had to hear many insults, and she received threatening telephone calls. Her family's home was even spray painted with obscene words.[44] Faulkner later admitted that she "had no idea" what she was getting into.[45] Talking about her experience at The Citadel she said, "I've been denied quite a few things—like a normal college life."[46]

South Carolina Institute for Leadership

On April 13, 1995, the Fourth Circuit ruled that Faulkner could join the cadet corps unless South Carolina had a similar leadership program for women, approved by the court by August. At this time, The Citadel began to explore options similar to VMI's women's leadership program. The military college offered to spend as much as $5 million to create a women's leadership program in South Carolina. The state had two private women's colleges—Columbia College and Converse College. Converse, located about two hundred miles from The Citadel in Spartanburg, was the only school interested in cooperating with The Citadel. On May 18, 1995, these two institutions agreed to each invest $5 million to create a leadership program at Converse. The program, called the South Carolina Institute for Leadership (SCIL), attracted twenty-two women to its first class. Sandra C. Thomas, president of Converse, believed this program would advance the rights of women in her state. She said, "The state has been providing money to men for years for single-gender education, and not for women. This would be a significant opportunity for women that does not exist."[47]

However, Marcia D. Greenberger, president of the National Women's Law Center, criticized The Citadel's program. She said she was not surprised that the South Carolina college was trying to implement a program similar to VMI's proposal. "The V.M.I. decision [encouraged] efforts to come up with the most minimal, face-saving methods that a court might accept based on stereotype."[48] A lawyer for Shannon Faulkner, Robert Black, agreed. He said the Converse plan would create "a powder-puff leadership program."[49]

In fact, the women's program at Converse was very different from The Citadel's program. Mealtime was just one example. At the military academy, knobs, wearing starched gray uniforms, sit in the brace position. Although they may sit at the table, their main job is to keep upper-class cadets' glasses filled with water, even as the older cadets yell at and insult them. The knobs occasionally get permission to eat a bite of food, but they must keep their chins tucked. Some knobs leave the mess hall as hungry as they went in.[50] At Converse College, women in the SCIL program wear civilian clothes and eat in a dining hall decorated with crystal chandeliers and floral curtains.[51] There were other differences between the harsh discipline of The Citadel and the kinder, gentler SCIL program. The women drilled, marched, and took ROTC training at nearby Wofford College. They had their own version of stressful training during a four-day Outward Bound program. Yet the SCIL program, unlike The Citadel's training, had few strict rules and no harsh discipline. This was because, in part, there were no upper-class cadets to enforce the rules and discipline.[52]

Yet the real difference between the programs at The Citadel and SCIL was in their philosophies. SCIL, like VWIL, used positive reinforcement and nurturing as opposed to stressful challenges and constant harass-ment. The interim director for SCIL at the time, Charles Gnerlich, said, "We take a woman who says, 'I can't do math or calculus,' and we give her a sense of confidence that she can achieve anything."[53] President Thomas of Converse believed that even if the methods were not the same, the outcomes would be comparable. She added, "Sometimes men need to be beaten down to get a concept. A woman gets it the first time."[54]

Echoing this viewpoint of the nurturing approach to women's education was Heather Wilson, the dean of students at Mary Baldwin College. She said,

> The VMI model is based on the premise that young men come with inflated sense of self-efficacy that must be knocked down and rebuilt in a more meaningful way. We believe that women had that leveling experi-ence already in their lives and they do not need more of that in college.[55]

Legal Battles Continue

As a result of her legal battles, Faulkner was finally admitted to The Citadel. On August 12, 1995, she reported to the military college with the other male cadets. Faulkner's parents as well as federal marshals accompanied her onto the college campus. During the second day of the knobs' stressful training routine," Faulkner became sick from exercising in one hundred-degree heat. Many saw her illness as a sign that women did not belong in military schools.

After four days in the infirmary, Faulkner quit The Citadel, because of the stress she felt from the challenging physical conditions and the hostility of the male students and their supporters. Attorney Robert Black described the vicious criticism Faulkner faced once she sued The Citadel. This harassment included T-shirts and bumper stickers throughout South Carolina that mocked her—frequently with obscene words. She also received nasty letters from people who did not want her admitted to The Citadel.[56] Faulkner herself said that the stress of two and a half years "came crashing down" on her, making her unable to remain part of the corps.[57]

Colonel Terrence E. Leedom, a spokesperson for The Citadel, denied that the school or its supporters had harassed Faulkner. "We went to great lengths to protect her," he said. "At the risk of saying 'I told you

Colonel Terrence E. Leedom, a spokesperson for The Citadel, denied that the school or its supporters had harassed Shannon Faulkner.

so,' this points out that the need for physical conditioning is obvious."[58] The Citadel claimed that Faulkner was overweight and out of shape and should not have been able to enroll in the corps.[59]

Legal experts did not think Faulkner's withdrawal from The Citadel would affect the goal of getting women admitted to the all-male school. "As far as the legal issue goes, it doesn't change anything except that it means The Citadel isn't going to be integrated right now," said Deborah Brake, a staff counsel at the National Women's Law Center.[60] Colonel Leedom agreed with Brake. "We'll now be facing the government, which never just represented Shannon Faulkner," he said.[61]

Perhaps the biggest impact of Faulkner's withdrawal was on public opinion. Many people viewed this as evidence that women do not belong in military schools. Deborah Brake said,

> I think there's a definite danger with the American public of people generalizing from this and not thinking about how hard it was to be the first woman to integrate that kind of institution, where there were so many people who wanted to see her fail.[62]

Brake believed the challenge would be to remind people that Faulkner's failure had no legal impact except to her own candidacy to enroll. "Under our equal-protection principles, you don't judge all women based on one woman's experiences," she said. "That's exactly the kind of stereotype that is not permitted."[63]

Case Sent to United States Supreme Court

In 1995, when the United States Court of Appeals for the Fourth Circuit ruled that VMI could exclude women from its program as long as it provided a similar program for women elsewhere, the United States government appealed. The Department of Justice claimed that Virginia was denying women equal protection under the law by operating a public institution for men without a similar program for women.[64] The government also argued that the appeals court relied on "harmful gender stereotypes" when it approved the VWIL program at Mary Baldwin College.[65]

After reviewing the government's appeal, the Supreme Court sided with the Department of Justice's viewpoint and reversed the 1995 decision of the court of appeals. The government's lawsuit against VMI was then sent to the Supreme Court for further study and review.

4

The Case for Admitting Women to VMI

Each year almost seven thousand cases are reviewed by the United States Supreme Court. The Supreme Court hears appeals from state courts and from district circuit courts if the case involves a federal question such as a constitutional issue. The nine Supreme Court Justices select fewer than one hundred cases for briefing and oral argument.[1] On January 17, 1996, the Supreme Court heard arguments on whether or not the male-only admissions policy of VMI was constitutional. The legal case was called *United States* v. *Virginia* and took five months to be decided.[2] The Supreme Court's decision in this case would impact The Citadel as well as VMI. Although the Justices would decide the question of admitting women to VMI on legal grounds, they spent quite a bit of time discussing social and philosophical issues. These issues included single-gender education and the adversative approach to education used by VMI and The Citadel.

Justice Stephen G. Breyer said that a desire to keep VMI's single-gender approach to education was not a good enough reason to discriminate against women. Using that type of thinking, he said, any school or college could bar members of a different group of people from attending.[3] Breyer asked Theodore B. Olson, the lawyer representing VMI, what was so important about the adversative method used at VMI that it could justify discriminating against women. Olson replied, "The answer, Justice Breyer, is that it works in a single-sex environment for men."[4]

In response to Olson's answer, Justice Ruth Bader Ginsburg asked, "Wouldn't something else work almost as well without denying anyone anything?"[5] Justice Ginsburg was referring to the VWIL program at Mary Baldwin College. In fact, she did not believe that the alternative women's program had the respect and tradition of VMI. She said that in many ways the program for women "does not qualify as VMI's equal."[6]

Deputy Solicitor General Paul Bender represented the United States government in this case. He presented

On January 17, 1996, the Supreme Court of the United States heard arguments on whether or not the male-only admissions policy of VMI was constitutional. Justice Ruth Bader Ginsburg did not believe the VWIL program was equal to that of VMI.

many of the plaintiff's points that were argued in earlier lawsuits against VMI. The government had five basic arguments for why women *should* be allowed to enroll in VMI:

1. VMI's admissions policy violated equal protection under the law of the United States Constitution.

2. By law, schools that receive federal or state funds are forbidden to discriminate against applicants based on gender.

3. The VWIL program was not equal to VMI's program, and a worthy alternative program could not be developed.

4. The four United States military academies already admitted women.

5. VMI's admissions policy practiced gender discrimination and promoted gender stereotypes.

The government's main argument was that VMI's admissions policy violated equal protection under the law of the United States Constitution. According to the Department of Justice, by refusing to admit female cadets, VMI was violating the rights of women. The military college was denying women equal opportunity to the education, training, and benefits that it provided to men. Bender argued, "It's inappropriate to say to a

59

particular woman who says 'I want that training,' 'You can't have it because you're a woman.'"[7]

Bender then pointed out that a publicly funded school is not allowed to discriminate against applicants because of their gender.[8] According to the Department of Justice, VMI officials had to either admit women or stop accepting state funds to finance the military college. If VMI had been a private school, it would have stood a better chance of keeping its all-male tradition. To support this point, the government relied on the Supreme Court's prior ruling in *Mississippi University for Women* v. *Hogan* (1982). The Supreme Court's ruling in that case forbids gender discrimination in schools that receive federal or state funding. The United States government did not see the VWIL program at Mary Baldwin College as an acceptable solution to VMI's discriminatory admissions policy. The Department of Justice found the VWIL program to be unequal as well as separate.[9] As in earlier arguments, many people said that the school's long tradition and approach to education could not be duplicated. Women in the VWIL program had no access to VMI's unique adversative education, and they enjoyed none of VMI's tradition. In addition, VWIL graduates did not have the same connections and contacts that the male VMI graduates had.[10] The government argued that if women could not receive anywhere else an opportunity like the one VMI offered, then the military college was obligated to open its door to women.

Many people felt that VMI's resistance to admitting women would be better understood had not the four United States military academies already allowed women. For twenty years, female applicants who met the academies' requirements had been allowed to enroll. Within that time, they stopped being novelties and rose to high positions of leadership.

In the *Hogan* case, Justice Sandra Day O'Connor had held that the Mississippi University for Women's admissions policy was discriminatory because it tended to promote the stereotype that nursing was solely a woman's job.[11] In light of the *Hogan* ruling, VMI's admissions policy was discriminatory because it promoted the stereotype that only boys should grow up to be soldiers.[12] In a brief filed for the United States, Solicitor General Drew Days said,

> VMI's admissions policy has communicated a message that, in the eyes of the Commonwealth, women do not possess the qualities of self-discipline, ability to withstand stress and respect for hierarchy [chain of command] that are widely associated with VMI.[13]

In an earlier brief, the Department of Justice had rejected VMI's argument that women could not withstand the adversative educational approach. Lawyers pointed out that the history of gender discrimination in the United States has been full of official assumptions that women and men properly belong in separate circles and play different social roles in society, according to gender.[14]

At the time of the Supreme Court case, Bill Clinton was president. He and his administration held that gender discrimination was just as offensive as race discrimination and that separate single-gender schools were ultimately unequal.[15] In fact, the legal arguments against discrimination based on race and gender are very similar. In a lawsuit for the National Women's Law Center, Susan Deller Ross and Wendy W. Williams repeated the arguments used in hundreds of gender discrimination cases in the 1970s. They claimed that gender, like race, is a characteristic unrelated to ability. Women, like African Americans, have suffered from a long history of discrimination in the United States; and women, like African Americans, are not fully represented in the political process.[16] The government felt that it was time for such open discrimination to end.

Overall, the United States Department of Justice believed that VMI needed to change—either by admitting women or by becoming a private institution. However, VMI officials, alumni, and cadets, as well as their supporters, had their reasons for resisting.

5

VMI's Case Against Admitting Women

Theodore B. Olson argued VMI's position in *United States* v. *Virginia.* He asked the United States Supreme Court to declare the school's admissions policy constitutional, even though women were barred from entering the military college. Like the government's position, VMI's position included many of the points argued in previous lawsuits challenging the school's admissions policy. The military college had the following five basic arguments for why women *should not* be allowed to enroll in the school:

1. Admitting women would destroy VMI's adversative educational approach and distract the male cadets.

2. By being one of only a few all-male schools, VMI offered diversity to Virginia's educational facilities.

3. The VWIL program at Mary Baldwin College was a valid alternative to VMI's program.

4. Women had plenty of other opportunities to receive military training.

5. Admitting women denies the physical and emotional differences between women.

VMI's main argument was that admitting women would destroy the military college's basic character and adversative tradition. Even before the United States Department of Justice sued VMI in 1990, the school and the state of Virginia had carefully reviewed the 1982 *Hogan* decision. In this earlier case of discrimination, the Supreme Court found fault with the Mississippi University for Women for not having revised its charter since its founding in 1884.[1] Before ever being sued for discriminatory practices, VMI actually took steps to review its charter.

In 1986, VMI had appointed a committee to study the school's mission and admissions policy. As part of the study, members of the committee visited the Army and Navy military academies to see how the admission of women impacted these schools. Committee members also met with supporters of coeducation. As a result of the findings from this lengthy study, VMI recommended that the school not change its male-only admissions policy. And in the 1991 federal court trial of the Department of Justice lawsuit, a district judge found that the 1986 report was a "reasoned analysis" of VMI's mission and its single-gender policy.[2]

The mission of VMI is to develop educated men of character who are honorable, self-disciplined, persistent, and loyal. The school's approach to education involves all areas of college life—living, studying, and playing. At most colleges the library is the main educational facility, but at VMI the Barracks are the focus of the school's educational program. Such an approach that includes all aspects of college life has produced graduates who are successful beyond expectations. Many people believed that by admitting women and changing the program VMI's educational system would be greatly harmed.

Major General Josiah Bunting, III summed up this point on behalf of VMI. The school's superintendent said, "If there is a college in the country whose essence [very being] is related to its gender, this is it. The changes would be seismic, profound, if we were to go co-ed."[3] Bunting denied that the military school was promoting stereotypes of women. He claimed that the

VMI's superintendent, Major General Josiah Bunting, III, summed up one argument for maintaining the all-male admissions policy.

exclusion of women from the military college was not based on the belief that being a soldier was solely man's work. He argued that VMI wanted to exclude women simply to preserve its unique, and successful, educational method for men.[4]

Many others at The Citadel agreed with Bunting. They believed that men learn better when they have to earn privileges and form bonds. The admission of women would distract the male cadets and interfere with the bonding process. Colonel Terrence Leedom said, "The showers, the haircuts, the lack of privacy all create bonding. If women entered here, the experience would be different."[5] Another important defense for VMI was the idea of diversity. VMI believed that its male-only policy served legitimate state interests, which would be compromised if the school were forced to admit women.[6] Even as early as 1991, the school's lawyers claimed that VMI complemented a system of public universities in the state of Virginia. They argued that segregation by gender was a logical consequence of diverse education.[7]

In earlier lawsuits, VMI lawyers argued that the military college's male-only admissions policy served lawful state interests. They said that these state interests would be compromised, or hold less value, if the school were forced to admit females.[8] VMI lawyers continued to argue that the VMI program added diversity to Virginia's public colleges. Supporters of VMI's all-male tradition believed that segregation by gender was an important feature. They felt that if VMI lost its ability to exclude women, it would also lose its ability to produce anything very distinctive.[9]

On the other hand, the "diversity" argument can sometimes be used to hide discrimination. Still, many people believed that in a time when coeducation was the overwhelmingly norm, a small number of single-gender colleges was not a serious problem. Newspaper columnist Ellen Goodman admitted that VMI was "less a threat to women's rights than an anachronism [outdated custom]."[10]

VMI claimed that the VWIL program at Mary Baldwin College was an acceptable option to the military education at VMI, especially in light of differences between the two genders. Olson argued that ruling against VMI would prohibit single-gender education in Virginia and also threaten the existence of every private single-gender school that gets state or federal money to operate. He said, "If there is a finding that single-sex education violates the Constitution, it is difficult for me to see how the state can support private single-sex institutions."[11] Olson was referring to the numerous single-gender schools that already existed, some of which received government funding.

In addition, VMI lawyers pointed to the fact that women were allowed to enroll in the nation's four military academies as well as other colleges that offered military training. They argued that these schools presented women with plenty of opportunities to receive military training, and consequently, VMI could and should remain an all-male school.

Some people believed that although arguments against gender discrimination were sorely needed in the 1970s, in the 1990s such arguments were outdated. By the end of the twentieth century, women had broken many barriers in various areas of society. VMI officials did not believe that the school's admissions policy discriminated against women, especially when other similar opportunities were available to women.

VMI's purpose was—and still is—to treat all cadets in exactly the same way.[12] However, school officials noted that gender differences do exist. They felt that because of women's different physical characteristics, VMI's tough fitness standards would have to be lowered.[13] They also thought that since women need more privacy than men, the college's environment of constant scrutiny and discipline would have to change.[14] To deny these gender differences was to ignore the realities associated with accepting women. And to compromise these differences was to forever change exactly what made VMI such an uncommon educational experience.

The Supreme Court Justices had heard all of the arguments for both sides. All that was left was to cast their votes to reach a majority decision.

6

The Supreme Court's Decision

On June 26, 1996, the Supreme Court announced its decision in the case of *United States* v. *Virginia*. There was no simple answer to the question of admitting or refusing applications from women to a military college with more than one hundred fifty years of tradition as an all-male school. As the president of Mary Baldwin College, Cynthia Tyson, said several years before the Supreme Court hearing, "I wish this were a simple issue, but unfortunately it's a complex issue.[1] Yet the Court had agreed to hear the case, and now, by law, had to rule on this matter.

Ruling Goes Against VMI

In a vote of seven to one, the Justices of the Supreme Court ruled against the all-male admissions policy of the state-run VMI. The Court decided that VMI's policy discriminated against women and violated the Constitution's guarantee of equal protection under the law.[2] The Court also ruled that the VWIL alternative all-female program at Mary Baldwin College did not correct the constitutional violation. The Court's ruling overturned a 1995 appeals court decision that VMI could continue to admit only men if the state of Virginia created an alternative program for women. Although VMI was the only defendant in the Supreme Court case, the Court's decision also applied to The Citadel—the nation's only other all-male public college. Both military schools were now faced with the challenge of admitting women to their traditionally all-male schools or possibly of making their schools private institutions.

Three of the Justices, Chief Justice William Rehnquist, Justice John Paul Stevens, and Justice Sandra Day O'Connor had participated in the 1982 case *Mississippi University for Women* v. *Hogan*, and all voted with the majority. Justices Anthony Kennedy, David Souter, Ruth Bader Ginsburg, and Stephen Breyer also voted to strike down the VMI policy. Justice

Antonin Scalia cast the only vote in favor of the VMI policy. Justice Clarence Thomas did not participate in the case because his son, Jamal, was a senior at VMI at that time.

Justice Ginsburg announced the decision from the bench in slow firm tones. She said that generalizations about men or women cannot be allowed to deny opportunities to those who do not fit these profiles.[3] She also compared VMI's leadership program for women at Mary Baldwin College to a separate law school for African Americans.

> Virginia's VWIL solution is reminiscent of the remedy Texas proposed 50 years ago, in response to a state trial court's 1946 ruling. . . . Reluctant to admit African Americans to its flagship University of Texas Law School, the state set up a separate school for Heman Sweatt and other black law students. . . . The Court unanimously ruled that Texas had not shown substantial equality in the [separate] educational opportunities the state offered.[4]

Justice Ginsburg wrote for the majority in a forty-one-page opinion that dismissed VMI's claim that admitting women would hurt the institution. She said, "However liberally this plan serves the State's sons, it makes no provision whatever for her daughters. This is not equal protection."[5]

Some people believed that Ginsburg's opinion

reflected many of the contradictions and questions about feminism and women's equality. On one hand, Ginsburg held a strict view of equality, holding that men and women should be treated the same. She believed that any woman who desired to enter VMI and was physically and mentally fit to do so should be given the opportunity to enroll. She said,

> The United States maintains that the Constitution's equal protection guarantee precludes [prevents] Virginia from reserving exclusively to men the unique educational opportunities VMI affords.[6]

However, on the other hand, Ginsburg used gender stereotypes herself in her arguments. She cited the district court's assertions that admitting women to VMI would require changes in the school's living arrangements to maintain privacy and changes in its physical training programs to address gender differences in physical ability.[7]

Concurring Opinion

Chief Justice Rehnquist concurred (agreed) with the Court's conclusion but disagreed with its reasoning. For that reason, he wrote a separate opinion. Rehnquist believed that Virginia's fault was not in keeping women from VMI but offering an alternative to women that

In a vote of seven to one, the Justices of the United States Supreme Court ruled against the all male admissions policy of VMI. Chief Justice William Rehnquist concurred (agreed) with the Court's conclusion, but offered his own reasoning.

was "distinctly inferior to the existing men's institution and will continue to be for the foreseeable future."[8]

Dissenting Opinion

Justice Scalia offered the lone dissent. In his forty-page dissent he claimed that VMI would be ruined if forced to alter its adversative method of education and abandon its most-defining trait—its maleness. Scalia wrote,

> The tradition of having government-funded military schools for men only is as well-rooted in the traditions of this country as the tradition of sending only men into military combat. The people may decide to change the one tradition, like the other, through democratic processes; but the assertion that either tradition has been unconstitutional through the centuries is not law, but politics smuggled into law.[9]

Other supporters of the Supreme Court decision also made opposing statements. Judith Lichtman of the Women's Legal Defense Fund said that the Court's ruling "unequivocally [unmistakably] outlaws the use of gender-based stereotypes."[10] One week later Marcia D. Greenberger of the National Women's Law Center warned that any attempt to impose identical physical standards on males and females at VMI would be "in conflict with both the letter and spirit of the Supreme Court decision," and would be challenged in court.[11]

Supreme Court Justice Antonin Scalia offered the lone dissent (disagreeing opinion) in the case. He felt that VMI would be ruined if forced to alter its adversative method of education and to abandon its all-male tradition.

Although the Court's ruling focused on public colleges and universities, many private schools for men or women were afraid the decision would force them to integrate because they accepted government money or tax relief.[12] Justice Scalia fueled this worry by warning that the Court's ruling would threaten not just public single-gender colleges and schools, but private ones as well. He said, "The only hope for state-assisted single-sex private schools is that the Court will not apply in the future the principles of law it has applied today."[13]

Many legal experts disagreed with Scalia's dissent. They pointed out that Justice Ginsburg's opinion focused on public institutions, not private ones. Also, public single-gender schools could remain that way as long as states provided equal treatment to men and women.[14]

Still, the Court's decision had a great impact on VMI and Citadel supporters. According to VMI's superintendent, Josiah Bunting, III, the Court's decision was a "bitter, bitter blow" for VMI alumni, who helped pay the school's $14 million in legal bills.[15] Several officials at VMI and The Citadel expressed their disappointment with the ruling but said they would obey it. On June 28—in a letter to colleagues, parents, cadets, friends of VMI, and brother rats—Bunting wrote the following regarding the Supreme Court's decision:

> I fought against this outcome since the suit was first brought against VMI. I know that it is as disagreeable, even galling, to you as it is to me. But we must now "conscious of duty faithfully performed" go forward in accordance with the law of our country, as our highest tribunal has interpreted it. I particularly expect gentlemen of the Corps to act and speak with dignity, forbearance and with continuing solicitude [concern] for the reputation of our noble and beloved Institute.[16]

Bunting said that nothing in the Supreme Court's opinions prevented VMI from becoming a private college.[17] However, he believed there was a good chance that the Department of Justice would block such a move, viewing it as an attempt to disobey the Court's order. For just as the federal government stopped schools in the 1950s and 1960s from closing their doors to African-American students by going private, it most likely would stop VMI from going private now that the school was found to discriminate against women.[18]

Also, making VMI a private school would be very expensive, costing from $100 million to $400 million.[19] The college sits on public land and receives about a third of its yearly budget from the state.[20] So the college would need a large amount of money, raised mainly by donations and by increasing tuition and room-and-board fees. According to Bunting, even if VMI chose to stay a public school, it would not be able to admit women until the fall of 1997.[21]

In the months following the Supreme Court's ruling, VMI's Board of Visitors voted to admit women to the school for the first time in the college's 157-year history. The vote was very close—nine to eight. "As you can see by the closeness of the vote, this resulted from a thorough and often difficult debate," said the board's president, William H. Berry. "This is not a decision we made easily, but we shall welcome the women who come here ready to meet the rigorous [difficult] challenges that produce the nation's finest citizen-soldiers."[22]

Colonel Michael M. Strickler, a VMI spokesperson, pointed out an interesting fact. Of the alumni on the Board of Visitors, those who had graduated before the 1970s tended to vote for admitting women. Younger graduates voted to make the school private rather than admit women.[23] It was decided that the female cadets would be admitted beginning in the fall semester of 1997.

General Bunting pointed out that recruiting women was just the first of many challenges now facing VMI.[24] The college was going to have to determine how and where to house the women, which new athletic programs to make available to women, and how to handle the awarding of scholarships that were marked for young men. Bunting predicted that VMI would be changed in ways that couldn't be imagined at that time.[25]

In fact, many felt that the presence of women would so change VMI as to deny all its students the type of distinctive education many enrolled to get. As Jeffrey Rosen said in *The New Republic*, the tiny minority of women who might come to a coeducational VMI "will not be able to achieve the precise benefits they seek," because "the institute in its current form will no longer exist."[26] In other words, women—by their very presence at the school—would change the school, so they would not get the same education that cadets previously received at the all-male school.

VMI would take a year to prepare for the admission of women to its campus. The preparation process included conducting seminars with cadets to explain what was expected of them, recruiting women from other military colleges, bringing in psychologists to help with the emotional transition, and hiring a public relations firm to handle the media.

The Citadel, on the other hand, began admitting women almost immediately. Two days after the Court's ruling, the school's Board of Visitors voted unanimously to accept women into the Corps of Cadets. This decision came after just two hours of private discussion by the board. In a public statement, James E. Jones, Jr., chairman of The Citadel's Board of Visitors, said, "The Citadel will enthusiastically accept qualified female applicants into the Corps of Cadets."[27] The South Carolina military college opened its doors to women in August 1996.

After the long legal battles and years of heated debate, the American public would get a chance to see the real impact of admitting women to these traditionally all-male military colleges.

7

Impact of the Court's Decision

Naturally, the United States Supreme Court's decision in *United States* v. *Virginia* had a huge impact on VMI and The Citadel. However, it also affected American society in general. The decision opened further debate about the issues of single-gender education and the meaning of true equality. And it raised many questions about how to integrate women into the all-male military schools.

First, the Court's ruling refueled the debate about private single-gender schools in the United States, most of which are all-female facilities. Are these educational institutions constitutional? Should any single-gender facility be allowed to exist in a country that believes in equal opportunity and treatment for all its citizens? Should private single-gender schools receive any government funding when they provide education to just one segment of the country's population?

One recent example of this debate is the Young Women's Leadership School in East Harlem, New York. The school was founded by Ann Rubenstein Tisch to help minority girls reach their dreams and goals. The school is now under attack by feminists and other people with liberal political beliefs who have filed a civil rights challenge against the school.[1] These people do not believe that an all-girls school offers fair treatment to boys. They also do not believe that it is good educational policy to teach children in single-gender schools. They think that the best learning occurs in integrated schools, where children learn and work together in an environment that includes students of different genders and religious or ethnic backgrounds.

The *VMI* case also left open the extent of gender equality. Does equal opportunity mean equal responsibility and equal requirements? Can young women be given the equal opportunity to enter VMI or The Citadel and yet not be held to the same physical standards to which the young men are held? Should they have to endure the same stressful situations that the male cadets do? In a footnote to the Supreme Court's opinion it was acknowledged that

> admitting women to VMI would undoubtedly require alterations [changes] necessary to afford members of each sex privacy from the other sex in living arrangements, and to adjust aspects of the physical training programs.[2]

Later the Court listed seventy-nine issues that the colleges had to address in integrating women, including sexual harassment policies and the uniforms of female cadets.[3]

Many of the issues raised by the Court's ruling are still being discussed. However, after the Supreme Court's 1996 ruling in *United States* v. *Virginia*, VMI and The Citadel were forced to change their admissions policies, which had previously allowed only men to enroll. Each school had its own approach to dealing with the Court's order.

VMI Starts to Comply With the Court's Decision

In the fall of 1996, VMI mailed about seventy applications to female high school seniors who requested information about entering the college. Major General Josiah Bunting, III, the school's superintendent, formed several panels to study the integration of female cadets. "Changes contemplated [considered] to accommodate the enrollment of women will be absolutely minimal [the least possible]," he said.

> Female cadets will be treated precisely as we treat male cadets. I believe fully qualified women would themselves feel demeaned [put down] by any relaxation in the standards that the V.M.I. system imposes on young men.[4]

General Bunting believed that VMI's approach to integrating women was within the boundaries of the Supreme Court's ruling. The Court's decision allowed that some women could meet the physical standards now imposed on men. The Court stated that neither VMI's goal of producing citizen-soldiers nor the school's method of doing so was inherently (by nature) unsuitable to women.[5] United States District Court Judge Jackson L. Kiser was put in charge of overseeing VMI's integration of women cadets. And by the end of the year a federal judge ruled that VMI had to report four times a year on its progress in integrating women.[6]

VMI decided to hold its female cadets to the same fitness standards as the male cadets. The physical tests require five pull-ups, sixty sit-ups in two minutes, and a mile-and-a-half run in twelve minutes.[7] Critics of VMI's policy said that holding female cadets to the same fitness standards as male cadets would discourage women from applying. Samuel B. Witt, III, a member of the VMI Board of Visitors and a 1958 VMI graduate—who voted to admit women—strongly disagreed. He said, "People are saying we're trying to discourage women, but I don't think the women who would be interested in V.M.I. would be discouraged."[8]

The Citadel Is First to Admit Women

In July 1996, The Citadel sent letters to nearly two hundred women who had expressed interest in attending the military college during the past few months. The school also began processing the five applications it had already received from women. Of those five, two applicants had already been notified that they met all admissions requirements and were eligible to enter the school in the fall semester.[9]

A twenty-page policy manual addressed the changes that would be necessary to integrate women into The Citadel's corps of cadets. The manual included information on issues such as uniforms, makeup, bathroom accommodations, and dating.[10] Some rules were as follows: Upper-class female cadets may wear transparent nail polish and a little makeup, but no jewelry may be worn by knobs; upper-class female and male cadets may date as long as they are not members of the same chain of command; male cadets must wear bathrobes when not otherwise clothed; cadets may not engage in sexual activity while on campus.[11]

The female cadets were to be disciplined under the Fourth-Class System, but were protected from sexual harassment. The women could not be touched without permission. The Citadel decided to hold its female cadets to the United States Army's physical-fitness standards for women. School administrators also permitted the female cadets to wear very short hairstyles rather than have their heads shaved in the style of the male cadets.

Attorneys for the women seeking admission to The Citadel voiced some criticisms of the school's integration plans. One issue was that of pregnancy. The attorneys noted that "while the defendants [Citadel officials] propose to dismiss pregnant cadets, they are silent as to the consequences to any male cadet who is found to be the father."[12] This and other issues were discussed at a hearing. Val Vojdik, a lawyer and longtime advocate to integrate the military college, said, "It's not enough to just open the door to women. You need to look behind the door and make sure they are being treated fairly."[13]

In August 1996, four young women entered The Citadel, along with about 570 male knobs. Plenty of media representatives were on hand to record the historic event. Perhaps for the first time ever in the college's history, a school official, Chaplain Charles Clanton, used the term "a Citadel woman" in the same breath as "a Citadel man."[14] A senior at the school, Cadet Jay Pearcy, said, "Ten years from now when we come back to this school, I'm going to be proud. Our class gets to make this [history]."[15]

The female knobs were Jeanie Mentavlos, Nancy Mace, Kim Messer, and Petra Lovetinska. The women reported to Barracks 2, which had locks installed on its doors, at different times throughout that Saturday morning. Each of the women was helped with her bags, escorted to her room, and introduced to members of the Cadet Corps, who for the next several months would lead them through the Citadel's traditional Fourth-Class System.[16]

Like all knobs, the women were kept away from the media. However, some family members spoke to the press. Petra Lovetinska's father, Jaroslav, an immigrant from Czechoslovakia, had his words interpreted by his son. He said that Petra wanted to go to The Citadel because it is a very good school. She had been in ROTC and her grandfather had been in the Czechoslovakian Army.[17] Citadel Cadet Michael Mentavlos commented on his sister's reason for entering the military college.

"My sister is not here to prove any points or gain attention. She's taking advantage of an opportunity."[18]

After reporting to the Barracks and changing into their blue physical fitness uniforms, the knobs had lunch in the mess hall with their parents. Then they attended several orientation talks and a religious gathering in the chapel. After a final good-bye to their family members, the women marched alongside the men, headed for their life as Citadel knobs.[19]

Throughout the next several months, the media received regular reports about how the female cadets were doing. The message was that the women were doing fine. However, by the end of the first semester, the public would learn otherwise.[20]

The Citadel Faces Problems

By the semester break in December 1996, 77 out of 581 knobs had departed The Citadel, which was not unusual.[21] However, in January 1997, The Citadel faced the extraordinary—sexual harassment charges. Two of the four female cadets, Jeanie Mentavlos and Kim Messer, announced that they were quitting the Citadel because of unfair and abusive treatment.

Among other instances of unfair treatment, the women said that upper-class cadets banged on their doors at all hours of the night. Then, and at other

In January 1997, Jeanie Mentavlos, one of four female cadets at The Citadel, announced she was quitting the military college because of unfair and abusive treatment.

times, the male students would sing obscene songs to them. They said cadets purposely used crude language in their presence, forced them to drink alcohol, and washed out their mouths with cleanser. Mentavlos said that at one time a male cadet rubbed his body against hers as she stood in line.[22]

What perhaps became the most publicized act against the women was an incident in which upper-class cadets poured polish remover on the sweat suits of two female knobs. Then the polish remover was ignited, setting the young women's clothing on fire. Cadets claim that this is a longstanding method of hazing knobs. Mentavlos said that after beating out the flames twice, one of the two cadet sergeants said, "Light her up again."[23] Mentavlos also claimed that once a cadet warned her, "If I ever see you off campus, I'll cut your heart out."[24]

In a prepared statement to the press Messer said, "I never asked for special treatment at The Citadel." Yet that is exactly what she felt she received, "special treatment by way of criminal assaults, sadistic [cruel] illegal hazing and disgusting incidents of sexual harassment."[25] Both Mentavlos and Messer filed lawsuits against the military college.

Shannon Faulkner, who had left The Citadel a year and a half earlier than this incident, commented on the

Kim Messer was another one of the four female cadets at The Citadel who announced in January 1997 that she was quitting because of unfair and abusive treatment.

women's accusations against the military school. At the time, she was attending Furman University in South Carolina. She said the harassment charges confirmed what she had already come to believe. She believed that The Citadel was not going to change its ways very soon. "For so long, I never said a negative word about The Citadel. But The Citadel that I believed existed does not exist."[26]

Some critics of The Citadel echoed Faulkner's feelings. Attorney Vojdik said, "A huge part of the problem is that you have the same administration that fought tooth and nail to keep women out, and now they're responsible for integrating it. That will just not work."[27]

However, it is not clear that these two young women were treated more harshly than the other first-year students simply because of their gender. The clothing of two male knobs was also set on fire with polish remover. In addition, the women were thought to have been targeted for extra hazing because they both had been excused from some duties back in September. Both Messer and Mentavlos had suffered pelvic stress fractures that prohibited them from participating in marching and running exercises. Pelvic stress fractures are common among women in the military, but the injuries may have been the "weakness" that some cadets were looking to exploit.

In a 1980 evaluation of the Fourth-Class System, evidence indicated that cadets with illnesses or temporary disabilities have been traditionally harassed, mistreated, and disrespected.[28] The acting president for The Citadel, Major General Roger C. Poole, acknowledged this custom, saying, "It's human nature for cadets to resent anyone who isn't participating."[29]

On the positive side, the two other young women at military school seemed to be fitting in just fine with their fellow knobs. Nancy Mace, whose father was a graduate of The Citadel, attained a 3.7 average (out of 4.0 possible points) during the first semester. She also bested 145 of 150 cadets in a two-mile run.[30] Her mother commented, "Nancy says, 'If you're going to be part of the system, you've got to be part of the system.'"[31] Both Mace and Lovetinska finished their first year at the Citadel and returned the following year.

Petra Lovetinska felt the secret to succeeding at The Citadel was "believing in what you're doing and doing it with full force. Even if I wasn't the fastest runner in the company, I was there, trying hard. And I didn't try to draw attention to myself."[32] She recounted the feelings of some of the male cadets. "They told me, 'I don't like women at the Citadel, but when I see the way you are working, I don't have a problem with it.'"[33]

Nancy Mace was one of the first female cadets at The Citadel. In May 1999, she became the first woman to graduate from The Citadel.

The Citadel Regroups

Citadel officials insisted that they were doing everything possible to wipe out sexual harassment at the college. Judge C. Weston Houck of the United States District Court also said he would do everything he could to ensure that the two female Citadel cadets who were harassed would be safe if they decided to return. However, the women no longer felt welcome at the college and stuck to their decision to leave.[34] In addition, Michael Mentavlos, a senior honor student, dropped out of the college in support of his sister. He was just a few credits short of graduation but was eligible to receive a Citadel degree after earning his remaining needed credits elsewhere.[35]

After Messer and Mentavlos made their harassment charges, eleven cadets were brought up on disciplinary charges for their part in harassing the female cadets. Two of these cadets were suspended from the school. Judge Houck also decided to appoint an outside specialist to help determine whether the school's integration of women was proceeding as it should.[36] In addition, in the interest of the women's safety, panic buttons were installed in the rooms of the female cadets.

General Poole offered this view,

Petra Lovetinska, who transferred into The Citadel, felt that the secret to succeeding there was "believing in what you're doing and doing it with full force."

> I think what happened is not an indication that the
> majority of people don't want women here. I honestly
> believe that the vast majority of people associated with
> this institution want to make gender integration a
> success.[37]

Yet some people worried that The Citadel had a serious
problem. And that problem was the climate of fear and
humiliation that affected all new cadets.

One Citadel graduate said,

> The fundamental problem is that the school sees its
> system and its toughness as its reason for being. So
> when you come and start talking about reforming
> that, you're challenging everything that they think dis-
> tinguishes them from the rest of the world.[38]

In fact, Citadel officials were beginning to take a hard
look at the school's culture. They did not want to
encourage a system that isolated or punished the weak
or the different. Officials said that kind of culture had
little or nothing to do with building leadership or
character.[39]

In 1997, twenty young women enrolled at The
Citadel along with 559 young men. In November, a
sexual encounter between a female first-year cadet and
an upper-class cadet was reported. This type of behavior is banned at The Citadel and was quickly
investigated. John Grinalds, president of the college,
reminded people that the "unhappy circumstance

involved only two cadets. We have 1,750 other cadets who are moving ahead and succeeding in all aspects of cadet life."[40] Even before this incident, Citadel alumni were beginning to speak out about the military college not being properly prepared to accept women into the cadet corps.[41] To improve coeducation, it was suggested that The Citadel spend more time on sexual harassment and sensitivity training.[42] The Citadel also learned from VMI, which had finally opened its door to women.

VMI Enrolls Women in Its Cadet Corps

In August 1997, for the first time in the military school's history, women entered VMI's traditionally all-male corps. Thirty-one women—thirteen of whom hailed from Virginia—joined the ranks. Representatives from over seventy media organizations were on hand to record the actions of this first coed class.[43] Yet access to the new cadets was restricted. Reporters were at all times escorted by eleven cadets who were selected by the school's public relations department.[44]

Tom Warburton served as the top media relations cadet that year. He was responsible for representing VMI before the national press and for keeping a watch over everything that went on in the Barracks. He felt that the biggest worry was that, there will be someone who chooses to do something outside of protocol [the rules] and draw the attention of the outside world. We've got to make it clear that action will be decisive. His name and VMI's name will be completely separated.[45]

To help prepare for the onslaught of public attention, VMI turned to Douglas Hearle, a New York public relations expert who assists clients in handling problems during a crisis. Hearle was actually hired as a consultant by VMI six years before when the first discrimination lawsuit was filed against the school.[46] VMI also recruited female advisers from other military-oriented schools, such as Texas A&M. These advisers counseled the incoming women, a strategy that The Citadel decided to use at its school in 1997.

Other ways that VMI prepared for the admission of women were discussion meetings with the corps, sexual harassment training, and communication of general rules governing the integration of women. For an entire day, VMI cadets received six hours of training to explain how the institution would change. In addition, cadets were given stern lectures on policies dealing with sexual assault and harassment.[47]

In the past, upper-class cadets were allowed to open any rat's door without knocking. Once women were admitted, however, they had to determine if the room housed a male or female cadet. If the room was that of a female, and the shade was pulled down, cadets had to knock before entering. To help the military college adapt its facilities to female cadets, the state of Virginia gave VMI $5.1 million to renovate the Barracks. The money was also spent on installing emergency lights and telephones across the campus. In addition, new rules about dating had to be drawn up and communicated. First-year cadets may not date while undergoing the rigors of the rat system. After that, a cadet may date anyone in his or her own class, but not anyone who is an upper-class cadet.[48]

One week after starting the rat line, "just one woman—along with 13 men—quit VMI."[49] To the school and its supporters, the uneventful first week did not just happen by luck. It was the result of all the preparation that VMI undertook before admitting the first female cadets in its history.[50] Those who had been on the opposite side of VMI during the lawsuits voiced caution in judging the school's integration at such an early stage. Marcia D. Greenberger, copresident of the National Women's Law Center, said, "VMI owes these young women far more than just the absence of abuse."[51] Yet even critics had to agree that the path VMI took toward admitting and protecting female cadets was much better than the one taken by The Citadel.[52]

What most everyone agreed on was the value of time.[53] While The Citadel had just a few months to prepare for the arrival of women, VMI had just over a year to prepare. The Citadel began to borrow some of VMI's strategies to help it better integrate the female and male cadets. It also conducted sensitivity training for all students. In addition, key leaders from the military college were sent to the Marine Corps training center at Paris Island in South Carolina to observe and learn how the Marines train women.[54] And finally, The Citadel shortened the time first-year students spent in the knob system by one month. The school also banned an intense last day of the year when first-year students were deprived of sleep and stretched to their mental and physical limits.[55]

The class of 1998 was the last all-male class of The Citadel. Nancy Mace, who entered The Citadel with thirty college credits, graduated in May 1999. Mace made history for women and the college as well as herself, even though she told a reporter, "The changes have not been made because of women, but to improve the system."[56] Petra Lovetinska and ten other female cadets who transferred into The Citadel are scheduled to graduate in 2000.

In May 1999, Chih-Yuan Ho became the first woman to earn a degree from VMI. Later in that historic graduation ceremony, Melissa Kay Graham also received a VMI diploma. The two women, who entered VMI in August 1997 as transfer students, graduated to loud cheers from over a dozen female cadets in the bleachers.[57] After 160 years, VMI alumni now included women.

However, the integration of women into the previously all-male military schools is not yet complete. In fact, it may take several more years before the total effect of admitting women can be assessed. What is known is that adjustments to each school had to be made, and some problems arose from those changes.

Yet the integration process has not been entirely negative. The first women to graduate from these time-honored military colleges spoke well of their experiences. Nancy Mace said that her years at The Citadel helped build her character. "Once you leave the school, the military aspect is left behind, but the education stays forever. What we learned about honor and duty and teamwork will stay with us."[58] She continued, "In our society, I think we need leaders to uphold and respect people, and The Citadel is training women to take those positions."[59] Melissa Kay Graham noted that she once called her father and told him she wanted to leave the school. Now she said, "I'm more than glad I stayed."[60]

Colonel Michael Strickler said that the VMI cadets are doing a "super job" integrating women into the ranks.

In truth, the real success of the integration process will be when female cadets at the college are no longer news. Integration will be complete when all first-year cadets, regardless of gender, are simply rats. And ultimate victory will be achieved when the female graduates of VMI and The Citadel can fulfill, like their male counterparts have been able to for over a century and a half, General Jackson's ambition for them: "Be all that you resolve to be."

In the meantime, Colonel Michael Strickler credited the VMI cadets with doing a "super job." He commented, "In the long run the cadets had to make it [the integration of women] work. It wouldn't have happened if they hadn't stepped up and done the job.[61]

Such positive leadership is encouraging. It is also encouraging that both VMI and The Citadel continue to emphasize responsibility, duty, and respect—in the strictest of terms—to their cadets. Therefore, there is no reason to believe that VMI and The Citadel will not become first-rate coed colleges as they have always been first-rate all-male schools.

Questions for Discussion

Both VMI and The Citadel graduated their first female cadets in May 1999. At VMI, the first four-year female cadet is scheduled to graduate in 2001—exactly one hundred years after George C. Marshall graduated from VMI. The road to these first female graduates was long and difficult. And the milestones were reached only after great human struggle. Solving social issues such as this one is never easy. And even after the highest court in the land makes a ruling, its decision still requires time for full acceptance and compliance by society.

The effects of the Supreme Court case *United States* v. *Virginia* are still being felt. And the issue still offers important questions for today's society. Think about the following questions. How would you have voted if you had been a Supreme Court Justice deciding this case?

1. Is it important for an educational system to include students with different backgrounds and who have different opinions? Why or why not?

2. What are the pros and cons of single-gender schools?

3. What are the pros and cons of coed schools?

4. Is an adversative environment a good one to learn in? Why or why not?

5. Were the VWIL and SCIL programs for women good alternatives to the all-male military colleges? Why or why not?

6. Should women serve in combat as soldiers? Why or why not?

7. Should women hold all of the jobs that men hold? Why or why not?

8. Should men and women have different physical requirements when performing the same jobs, such as working in the police force or the fire department?

9. What characteristics make a strong leader? Can both men and women possess these qualities?

10. How should people treat those who are different from them, such as those with a different racial background, religion, gender, or sexual preference?

11. Should women have been allowed to enter VMI and The Citadel? What is the reasoning behind your opinion?

12. What have you learned from reading about *United States* v. *Virginia*?

Chapter Notes

Chapter 1. Welcome to the "Rat" World

1. George Hackett and Mark Miller, "Manning the Barricades," *Newsweek*, March 26, 1990, p. 18.

2. Karen Hauppert, "Where the Boys Are," *Village Voice*, June 4, 1991, pp. 35–39.

3. Hackett and Miller, p. 18.

4. Author telephone interview of Michael M. Strickler, Director of Public Relations, VMI, January 25, 1999.

5. Ibid.

6. William G. Broaddus, "Is the VMI Decision an Omen or an Aberration?" *The Chronicle of Higher Education*, July 12, 1996, p. A48.

7. Ibid.

8. Hackett and Miller, p. 19.

9. Author telephone interview of Michael Strickler, January 25, 1999.

Chapter 2. The Start of a Long March

1. *Academic American Encyclopedia*, vol. 5 (Danbury, Conn.: Grolier Incorporated, 1992), p. 212.

2. *Brown v. Board of Education*, 347 U.S. 483, 495 (1954).

3. *Brown v. Board of Education*, 349 U.S. 294, 300 (1954).

4. Carol Hymowitz and Michael Weissman, *A History of Women in America* (New York: Bantam Books, 1978), p. 85.

5. Ibid.

6. Ibid., p. 94.

7. Ibid., p. 96.

8. Ibid., p. 156.

9. "Women Rank High in Naval Academy Class," *The New York Times*, May 27, 1999, p. A21.

10. Susan Dodge, "In the Corps of Cadets at North Georgia College, Men and Women Learn and Train Side by Side," *The Chronicle of Higher Education*, March 21, 1990, p. A34.

11. Ibid.

12. *Mississippi University for Women* v. *Hogan*, 458 U.S. 718 (1982).

13. Scott Jaschik, "Legal Challenges to Single-Sex Colleges Expected to Spread," *The Chronicle of Higher Education*, February 28, 1990, p. A26.

Chapter 3. The Legal Battle

1. Emily Mitchell, "The Thin Gray Line," *Time*, July 1, 1991, p. 66.

2. Jere Real, "The Last of the Old Corps?" *National Review*, August 6, 1990, p. 23.

3. Ibid.

4. Karen Hauppert, "Where the Boys Are," *Village Voice*, June 4, 1991, pp. 35–39.

5. Scott Jaschik, "Judge Rules That Virginia Military Institute's Policy of Admitting Only Men Does Not Violate Law," *The Chronicle of Higher Education*, June 26, 1991, p. A15.

6. Hauppert, pp. 35–39; Mitchell, p. 66.

7. *Mississippi University for Women* v. *Hogan*, 458 U.S. 718 (1982).

8. Jaschik, "Judge Rules That Virginia Military Institute's Policy of Admitting Only Men Does Not Violate Law," p. A19.

9. Scott Jaschik, "3 Education Researchers Played Key Role in Decision to Uphold All-Male Policy at Va. Military Institute," *The Chronicle of Higher Education*, July 3, 1991, p. A11.

10. Ibid.

11. Ibid., p. A13.

12. Ibid., p. A11.

13. Jaschik, "Judge Rules That Virginia Military Institute's Policy of Admitting Only Men Does Not Violate Law," p. A19.

14. Ibid.

15. Ibid.

16. Mitchell, p. 66.

17. Ibid.

18. Jim Zook, "Virginia Can Run All-Male Military College If It Creates Women's Program," *The Chronicle of Higher Education*, October 4, 1992, p. A25.

19. Scott Jaschik, "7 Women's Colleges Back VMI's Appeal to Retain All-Male Student Body," *The Chronicle of Higher Education*, April 7, 1993, p. A23.

20. Zook, p. A25.

21. Ibid.

22. Ibid.

23. Jaschik, "7 Women's Colleges Back VMI's Appeal to Retain All-Male Student Body," p. A21.

24. 508 U.S. 946, 124 L. Ed. 2d 651, 113 S. Ct. 2431.

25. Scott Jaschik, "High Court Deals Blow to Va. Military Institute by Declining to Hear Its Appeal to Stay All-Male," *The Chronicle of Higher Education*, June 2, 1993, p. A21.

25. Ibid.

26. Ibid.

27. Zook, p. A22.

28. Scott Jaschik, "All-Male VMI Proposes to Stay That Way by Aiding Women's Colleges," *The Chronicle of Higher Education*, October 6, 1993, p. A29.

29. Ibid.

30. Ibid.

31. Ibid.

32. Ibid.

33. Ibid.

34. Jaschik, "High Court Deals Blow to Va. Military Institute by Declining to Hear Its Appeal to Stay All-Male," p. A21.

35. Jaschik, "All-Male VMI Proposes to Stay That Way by Aiding Women's Colleges," p. A29.

36. Jaschik, "High Court Deals Blow to Va. Military Institute by Declining to Hear Its Appeal to Stay All-Male," p. A21.

37. Scott Jaschik, "Virginia Military Institute's All-Male Status Upheld by Appeals Court," *The Chronicle of Higher Education*, February 3, 1995, p. A28.

38. Ibid.

39. Ibid.

40. 4th District Court of Appeals, 44F. 3d 1229 (CA4 199), p. 1250.

41. Ibid.

42. Jaschik, "Virginia Military Institute's All-Male Status Upheld by Appeals Court," p. A28.

43. Ibid.

44. "Women Ease Into Citadel Life, Routine," *Charleston Post & Courier,* Charleston, S.C., August 25, 1996, p. A1.

45. "Shannon Faulkner," *People Weekly,* December 26, 1994, p. 58.

46. Ibid.

47. Scott Jaschik, "U.S. Seeks Reversal of Decision to Let VMI Stay All Male," *The Chronicle of Higher Education,* June 9, 1995, p. A27.

48. Scott Jaschik, "VMI Ruling Buoys Citadel in Its Battle to Remain All-Male," *The Chronicle of Higher Education,* February 10, 1995, p. A33.

49. Jaschik, "U.S. Seeks Reversal of Decision to Let VMI Stay All Male," p. A27.

50. Melinda Beck and Nina A. Biddle, "Separate, Not Equal," *Newsweek,* December 11, 1995, p. 86.

51. Ibid.

52. Ibid., p. 87.

53. Ibid.

54. Ibid.

55. Jeffrey Rosen, "Like Race, Like Gender?" *The New Republic*, February 19, 1996, p. 21.

56. Scott Jaschik, "Experts Predict Female Cadet's Withdrawal Will Not Affect Legal Case Against Citadel," *The Chronicle of Higher Education*, September 8, 1995, p. A44.

57. "Shannon Faulkner's Citadel Chronology," *The Detroit News*, August 19, 1995, <http://detnews.com/menu/stories/13948.htm> (October 18, 1999).

58. Jaschik, "Experts Predict Female Cadet's Withdrawal Will Not Affect Legal Case Against Citadel," p. A44.

59. Ibid.

60. Ibid.

61. Ibid.

62. Ibid.

63. Ibid.

64. Jaschik, "U.S. Seeks Reversal of Decision to Let VMI Stay All Male," p. A27.

65. Ibid.

Chapter 4. The Case for Admitting Women to VMI

1. Written comments of William K. Suter, Supreme Court Clerk, June 18, 1999.

2. *United States* v. *Virginia*, 518 U.S., L ed. 2d 735, 116 S. Ct.

3. Douglas Lederman, "Supreme Court Hears Arguments on VMI Admissions Policy," *The Chronicle of Higher Education*, January 26, 1996, p. A28.

4. Ibid.

5. Ibid.

6. Douglas Lederman, "Supreme Court Rejects VMI's Exclusion of Women," *The Chronicle of Higher Education*, July 5, 1996, p. A21.

7. Lederman, "Supreme Court Hears Arguments on VMI Admissions Policy," p. A28.

8. John Diconsiglio and Jennifer MacNair, ". . . And Justice for All," *Scholastic Update*, September 20, 1996, pp. 12–14.

9. D. Grier Stephenson, Jr., "The Future of Single-Sex Education," *USA Today*, January 1997, pp. 80–82.

10. Ibid.

11. *Mississippi University for Women* v. *Hogan*, 488 U.S. 718 (1982).

12. Jeffrey Rosen, "Like Race, Like Gender?" *The New Republic*, February 19, 1996, p. 21.

13. Ibid.

14. Scott Jaschik, "U.S. Seeks Reversal of Decision to Let VMI Stay All Male," *The Chronicle of Higher Education*, June 9, 1995, p. A27.

15. Rosen, p. 21.

16. Ibid.

Chapter 5. VMI's Case Against Admitting Women

1. William G. Broaddus, "Is the VMI Decision an Omen or an Aberration?" *The Chronicle of Higher Education*, July 12, 1996, p. A48.

2. Ibid.

3. Peter Schmidt, "Gloom Pervades VMI Campus as Cadets Learn Their Ranks May Soon Include Women," *The Chronicle of Higher Education*, July 5, 1996, p. A26.

4. Jeffrey Rosen, "Like Race, Like Gender?" *The New Republic*, February 19, 1996, p. 21.

5. Melinda Beck and Nina A. Biddle, "Separate, Not Equal," *Newsweek*, December 11, 1995, pp. 86–87.

6. Karen Hauppert, "Where the Boys Are," *Village Voice*, June 4, 1991, pp. 35–39.

7. "Not When the State Pays," *The Economist*, vol. 340, no. 7973, July 6, 1996, p. 31.

8. Hauppert, pp. 35–39.

9. Wilfred M. McClay, "Of 'Rats' and Women," *Commentary*, September 1996, p. 46.

10. Ibid.

11. Douglas Lederman, "Supreme Court Hears Arguments on VMI Admissions Policy," *The Chronicle of Higher Education*, January 26, 1996, p. A28.

12. Rosen, p. 21.

13. John Diconsiglio and Jennifer MacNair, ". . . And Justice for All," *Scholastic Update*, September 20, 1996, pp. 12–14.

14. Ibid.

Chapter 6. The Supreme Court's Decision

1. Scott Jaschik, "7 Women's Colleges Back VMI's Appeal to Retain All-Male Student Body," *The Chronicle of Higher Education*, April 7, 1993, p. A23.

2. Douglas Lederman, "Supreme Court Rejects VMI's Exclusion of Women," *The Chronicle of Higher Education,* July 5, 1996, p. A21.

3. Aaron Epstein, "Supreme Court Ruling Expected to Open VMI, The Citadel to Women," *Knight-Ridder/Tribune News Service,* June 26, 1996, p. 626.

4. *United States* v. *Virginia* 135 L Ed. 2nd 735, 763–764, (1996).

5. Ibid., p. 756.

6. Ibid., p. 743.

7. Ibid., p. 746.

8. Ibid., p. 771.

9. Ibid., p. 774.

10. Ibid., p. 746.

11. Ibid.

12. Epstein, p. 626.

13. *United States* v. *Virginia* 135 L Ed. 2nd 735, 792 (1996).

14. Lederman, p. A21.

15. Peter Schmidt, "Gloom Pervades VMI Campus as Cadets Learn Their Ranks May Soon Include Women," *The Chronicle of Higher Education,* July 5, 1996, p. A26.

16. "Letter from the Superintendent—Supreme Court Decision," VMI Web site, June 28, 1996, <http://www.vmi.edu/~pr/letter.htm> (October 18, 1999).

17. Schmidt, p. A26.

18. Lederman, p. A21.

19. Wilfred M. McClay, "Of 'Rats' and Women," *Commentary*, September 1996, p. 46.

20. Schmidt, p. A26.

21. Ibid.

22. Karla Haworth, "VMI Board Votes to Admit Women," *The Chronicle of Higher Education*, October 4, 1996, p. A30.

23. Ibid.

24. Schmidt, p. A26.

25. McClay, p. 46.

26. Ibid.

27. Douglas Lederman, "Citadel Moves to Admit Women, but VMI Considers Alternatives," *The Chronicle of Higher Education*, July 12, 1996, p. A28.

Chapter 7. Impact of the Court's Decision

1. Anemona Hartocollis, "Call Her an Advocate, Not an Aristocrat," *The New York Times*, December 1, 1998, p. B2.

2. Douglas Lederman, "Citadel Moves to Admit Women, but VMI Considers Other Alternatives," *The Chronicle of Higher Education*, July 12, 1996, p. A28.

3. "Report From the Trenches at The Citadel: It's Working!" *Women in Higher Education*, February 1999, p. 1.

4. Karla Haworth, "VMI Board Votes to Admit Women," *The Chronicle of Higher Education*, October 4, 1996, p. A30.

5. *United States* v. *Virginia* 135 L Ed. 2nd 735, 744 (1996).

6. Douglas Lederman, "Judge Orders Virginia to Report on Progress in Enrolling Women at VMI," *The Chronicle of Higher Education*, December 13, 1996, p. A34.

7. Haworth, p. A30.

8. Ibid.

9. Lederman, "Citadel Moves to Admit Women, but VMI Considers Other Alternatives, p. A28.

10. "Citadel Releases Co-Ed Plan," *Charleston Post & Courier*, Charleston, S.C., August 6, 1996, p. A1.

11. Ibid.

12. "Citadel Integration Plan Contested," *Charleston Post & Courier*, Charleston, S.C., August 10, 1996, p. A1.

13. Ibid.

14. "Women Ease Into Citadel Life, Routine," *Charleston Post & Courier*, Charleston, S.C., August 25, 1996, p. A1.

15. Ibid.

16. Ibid.

17. Ibid.

18. Ibid.

19. Ibid.

20. "Women at The Citadel: What Went Wrong?" *Charleston Post & Courier*, Charleston, S.C., March 9, 1997, p. A1.

21. Bill Hewitt and Don Sider, "Conduct Unbecoming," *People Weekly*, January 27, 1997, pp. 40–43.

22. Ibid.

23. Ibid.

24. Ibid.

25. Ibid.

26. Ibid.

27. Ibid.

28. "Women at The Citadel: What Went Wrong?" p. A1.

29. Hewitt and Sider, pp. 40–43.

30. "A Long, Tough March Into History," *People Weekly*, December 29, 1997–January 5, 1998, p. 163.

31. Hewitt and Sider, pp. 40–43.

32. "A Long, Tough March into History, p. 163.

33. Ibid.

34. "Female Cadets Leaving Citadel," *Charleston Post & Courier*, Charleston, S.C., January 13, 1997, p. A1.

35. Ibid.

36. Ibid.

37. Hewitt and Sider, pp. 40–43.

38. Ibid.

39. "Women at The Citadel: What Went Wrong?" p. A1.

40. "A Long, Tough March Into History," p. 163.

41. "Citadel Alumnus: We Weren't Ready," *Charleston Post & Courier*, Charleston, S.C., February 21, 1997, p. B1.

42. Ibid.

43. Ron Nixon, "Media Coverage Limited as First Coed Class Enters VMI," *The Roanoke Times*, Knight-Ridder/ Tribune Information Service electronic collection, August 18, 1997.

44. Ibid.

45. Alison Freehling, "VMI Cadet in the Spotlight as Women Join Ranks for First Time," *Newport News Daily Press,* Knight-Ridder/Tribune Information Service electronic collection, August 18, 1997.

46. Nixon.

47. Charles Pope, "VMI's Co-Educational Transition Runs Smoother Than Citadel's," *Knight-Ridder/Tribune* News Service, August 22, 1997.

48. Freehling.

49. Pope.

50. Ibid.

51. Ibid.

52. Ibid.

53. Ibid.

54. Ibid.

55. Ibid.

56. "Female Cadet Graduates at The Citadel, a First," *The New York Times,* May 9, 1999, p. 16.

57. "First Women Received Diplomas From V.M.I.," *The New York Times,* May 17, 1999, p. A14.

58. "Report From the Trenches at The Citadel: It's Working!" p. 3.

59. Ibid.

60. "First Women Receive Diplomas From V.M.I.," p. A14.

61. Author telephone interview of Michael M. Strickler, Director of Public Relations, VMI, January 25, 1999.

Glossary

abolition movement—Public action taken to ban slavery in the United States.

admissions policy—Standards that an educational facility requires for enrollment.

adversative—Challenging or difficult.

appeal—Asking a court with greater authority to reconsider the decision of a lower court.

Bill of Rights—The first ten amendments to the United States Constitution.

brief—A written document that states the arguments attorneys want to make in their lawsuits.

charter—A written document that empowers a facility or group of people.

coeducation—Teaching that includes both genders in the same classes.

compliance—The act of yielding to or obeying an order or command.

concur—Agreement with the opinion of the majority.

defendant—A person or party being sued.

Department of Justice—Part of the executive branch of the federal government that handles lawsuits in federal matters and interprets and enforces federal laws.

dissent—A disagreement with the opinion of the majority.

diversity—The condition of being different.

equal protection—A constitutional guarantee that no person or group of people will be created unfairly under the law.

feminism—The theory of political, economic, and social equality between men and women.

gender discrimination—Bias against members of the opposite gender.

hazing—Harassing by forcing unpleasant activities on someone.

immigrant—A person who comes from another country.

initiation—A ceremony that makes someone a member of a facility or group of people.

integration—The act of uniting two different groups of people.

majority opinion—The written decision of a court in which the greater number of deciding Justices agree on an issue.

panel—A group of people selected for some type of service or activity.

plaintiff—The person bringing the lawsuit and suing the defendant.

prosecution—The person or party conducting the legal proceedings against the defendant.

race discrimination—Bias against people of a different ethnic background.

ratification—Formal approval by a legal body.

Reserve Officer Training Corps (ROTC)—A program that trains high school and college students to serve as officers in the United States armed forces.

resolve—Declare or decide.

reveille—A wake-up signal given in the morning.

reverse discrimination—Bias against people who are not normally treated unfairly.

segregation—The act of separating two different groups of people.

seminars—Meetings for discussing and or giving information.

separate but equal doctrine—The idea that two different facilities or policies can offer fair treatment to all.

sexual harassment—Bothering someone repeatedly regarding that person's gender or sexual preference.

single-gender schools—Educational facilities that teach either one gender or the other.

suffrage movement—Public action to give women in the United States the right to vote.

Further Reading

Beck, Melinda, and Nina A. Biddle. "Separate, Not Equal." *Newsweek*, December 11, 1995, pp. 86–87.

Breuer, William B. *War and American Women: Heroism, Deeds and Controversy.* Westport, Conn.: Greenwood Publishing Group, Inc., 1997.

Bunting, Josiah III. "Making Room for Sister Rat." *Newsweek*, December 23, 1996, p. 54.

Diconsiglio, John, and Jennifer MacNair. " . . . And Justice for All." *Scholastic Update*, September 20, 1996, pp. 12–14.

Hackett, George, and Mark Miller. "Manning the Barricades." *Newsweek*, March 26, 1990, pp. 18–20.

Hewitt, Bill, and Don Sider. "Conduct Unbecoming." *People Weekly*, January 27, 1997, pp. 40–43.

Kaplan, David A. "VMI Braces for a Few Good Women." *Newsweek*, July 8, 1996, p. 72.

"A Long Tough March into History." *People Weekly*, December 29, 1997–January 5, 1998, p. 163.

Mitchell, Emily. "The Thin Gray Line." *Time*, July 1, 1991, p. 66.

"Shannon Faulkner." *People Weekly*. December 26, 1994, p. 58.

Skaine, Rosemarie. *Women at War: Gender Issues of Americans in Combat.* Jefferson, N.C.: McFarland & Company, Inc., 1998.

Stephensen, D. Grier, Jr. "The Future of Single-Sex Education." *USA Today*, January 1997, pp. 80–82.

Internet Addresses

The Citadel
  <http://www.citadel.edu>
Virginia Military Institute
  <http://www.vmi.edu>

125

Index

128